Chapter 1: The Science of Bioelectricity

Bioelectricity is a fundamental concept in biology, bridging the gap between the world of electrical engineering and the intricate cellular processes that govern life. It refers to the electrical phenomena that occur within living organisms, particularly the electrical signals that travel through the body to regulate various biological functions. Understanding bioelectricity is essential for harnessing the power of AI, RF therapy, and electromagnetic stimulation for muscle growth and regeneration.

Fundamentals of Bioelectricity

Bioelectricity begins at the cellular level. All living cells generate electrical potentials, which are typically called **membrane potentials**. These potentials are created by the difference in the concentration of ions inside and outside the cell membrane. Specifically, the balance of ions like sodium (Na^+), potassium (K^+), and chloride (Cl^-) creates a voltage difference across the cell membrane, which can vary in magnitude.

The resting membrane potential is typically around -70 millivolts (mV) for most cells, but this can change rapidly when a cell is stimulated. For example, neurons and muscle cells can undergo rapid changes in membrane potential to generate action potentials, the electrical signals that propagate along nerves or muscle fibers. These action potentials are crucial for transmitting information throughout the body, controlling everything from muscle contractions to brain function.

In muscle cells, electrical signals trigger the contraction process. **Muscle fibers** respond to these electrical impulses by contracting, leading to movement. The ability of cells to generate and respond to electrical signals is key to the functioning of the nervous system, the heart, and all other muscle tissues. This is where bioelectricity connects directly to **muscle regeneration**, as it is the same electrical mechanisms that govern muscle contractions and growth.

Electric Signals in the Body

The human body relies on a network of electrical signals to communicate internally, with **nerve impulses** being one of the most common forms of bioelectric communication. **Nerves**, particularly those involved in motor control, generate electrical signals (action potentials) that travel rapidly to muscles, instructing them to contract. These electrical impulses are generated by **ion channels** in the neuron's membrane. When a signal reaches a neuromuscular junction, it prompts the muscle to contract, starting a cascade of biochemical events that involve **calcium ion release** and the interaction of actin and myosin filaments, leading to muscle contraction.

In addition to muscle contractions, electrical signals regulate other crucial bodily functions, such as heart rhythms, brain activity, and organ function. The **electrical properties** of cells allow for rapid communication and synchronization of processes across vast distances in the body. This is evident in how the heart generates its electrical impulses, which control the rhythmic contractions of the heart muscle.

One critical concept in bioelectricity is **electrophysiology**, the study of the electrical properties of biological cells and tissues. This field helps explain how electrical signals propagate through tissues like nerves and muscles and how external electric fields can influence biological processes. The key to harnessing these properties for muscle regeneration lies in our ability to **manipulate electrical signals** to promote healing and growth in targeted tissues.

Electromagnetic Fields and Biological Tissue

Electromagnetic fields (EMFs), which are produced by both natural and man-made sources, have profound effects on biological tissues. These fields are made up of electric and magnetic components, and they can interact with the body's bioelectric systems in various ways. Understanding how EMFs influence biological tissues is crucial for developing therapies that use **radiofrequency (RF)** and **electromagnetic stimulation (EMS)** to promote muscle growth and regeneration.

For example, **radiofrequency fields**, which fall within the electromagnetic spectrum, are already used in medical treatments such as RF therapy. RF waves are electromagnetic waves with frequencies between 3 kHz and 300 GHz. When applied to tissues, RF energy can penetrate several millimeters deep, inducing thermal and non-thermal effects. These effects are useful in stimulating tissue regeneration, enhancing **collagen production**, increasing **blood circulation**, and promoting **cellular repair**.

One of the fundamental mechanisms by which EMFs affect biological tissue is **ion movement**. EMFs can influence the movement of ions in cells, altering their electrical charges and creating conditions that stimulate growth factors, cellular repair, and regeneration. When RF or electromagnetic fields are applied strategically, they can trigger biochemical reactions that lead to the activation of growth factors and hormones, such as **growth hormone (GH)**, which is vital for muscle repair and regeneration.

Furthermore, **electromagnetic stimulation** has been used for wound healing and tissue regeneration, which highlights its potential for promoting muscle growth. Studies have shown that when specific frequencies of electromagnetic fields are applied to injured tissues, they can accelerate the healing process by increasing the proliferation of **fibroblasts**, which are the cells responsible for producing collagen and other extracellular matrix components.

How RF and Electromagnetic Fields Affect Biological Tissues at the Cellular Level

At the cellular level, RF and electromagnetic fields influence a variety of physiological processes. For instance, RF therapy, which uses electromagnetic waves to generate heat in tissues, can stimulate the **mitochondria**, the powerhouses of the cell, to increase energy production. This boosts cellular metabolism and enhances the tissue's ability to repair itself.

In muscle cells, RF and electromagnetic stimulation can promote **muscle cell activation** by encouraging the release of **calcium ions** from the sarcoplasmic reticulum, which is crucial for muscle contraction. When applied appropriately, these signals can enhance the efficiency of muscle contractions, promote muscle fiber regeneration, and reduce recovery time.

Additionally, RF and electromagnetic fields can also influence **cell signaling pathways** that control inflammation, blood flow, and tissue healing. These pathways are essential for ensuring that muscle fibers are repaired and regenerated after exercise or injury. The fields can enhance **vascularity**, improving blood supply to the muscles, which increases nutrient delivery and accelerates the repair process.

The use of **electromagnetic stimulation** in clinical and therapeutic settings has grown in popularity due to its non-invasive nature and ability to target specific tissues. For example, **transcranial magnetic stimulation (TMS)** is used in brain therapies, and **pulsed electromagnetic field therapy (PEMF)** is used for bone healing and muscle regeneration.

By harnessing these electromagnetic properties, researchers and clinicians can accelerate the process of muscle growth, repair, and regeneration—essentially mimicking the natural processes of cellular growth but with the added power of AI and targeted electromagnetic fields.

The Path Forward

Understanding the fundamental science of bioelectricity is the first step in unlocking the potential of RF therapy and electromagnetic stimulation for muscle regeneration and growth. By exploring how electrical signals function within the body, we can begin to see the vast opportunities that lie in manipulating these signals to foster muscle repair, enhance growth hormone production, and optimize overall muscle health.

In the following chapters, we will delve deeper into how AI, RF, and electromagnetic stimulation are not just theoretical but practical technologies for revolutionizing muscle regeneration. We will also explore the ways in which these technologies are being used together to create personalized, AI-driven muscle growth systems, offering new hope for anyone looking to optimize their muscle health and recovery.

This chapter introduces the fundamental science of bioelectricity, setting the stage for exploring how electrical signals can be manipulated to enhance muscle growth and regeneration. Future chapters will build upon this knowledge to explore practical applications, especially in the context of advanced technologies like AI, RF therapy, and electromagnetic stimulation.

Chapter 2: Muscle Physiology and Growth

Muscle physiology is a complex and intricate process, and understanding it is key to optimizing muscle growth and regeneration. This chapter provides an in-depth exploration of the structure and function of muscle fibers, how muscle growth occurs, and the mechanisms behind muscle repair. It also delves into the role of **mechanical stress** and **nutrition** in the muscle regeneration process, laying the foundation for integrating bioelectric technologies such as **RF therapy** and **electromagnetic stimulation (EMS)**.

Muscle Structure

Muscle fibers are the basic building blocks of muscle tissue. The body contains three primary types of muscle: **skeletal**, **smooth**, and **cardiac**. Skeletal muscles, the focus of this chapter, are responsible for voluntary movements like walking, lifting, and running. These muscles are composed of long, cylindrical cells called **muscle fibers**.

Muscle fibers are unique because they are multinucleated, meaning they contain more than one nucleus per cell. Each muscle fiber is made up of smaller units called **myofibrils**, which contain repeating structures known as **sarcomeres**. Sarcomeres are the functional units of muscle contraction and are composed of two key proteins: **actin** (thin filaments) and **myosin** (thick filaments). These proteins interact during muscle contraction, pulling the actin filaments toward the center of the sarcomere, causing the muscle fiber to shorten and contract.

When a muscle contracts, the individual muscle fibers contract as well. The force generated by the contraction of millions of fibers is what powers movement. The structure of muscle fibers allows them to be highly adaptable, capable of responding to mechanical stress by increasing their size (hypertrophy) or number (hyperplasia).

The Role of Protein Synthesis in Muscle Growth

Muscle growth is fundamentally driven by the process of **protein synthesis**. After a workout or injury, the body initiates a process where damaged muscle fibers are repaired by synthesizing new proteins. These proteins, primarily **actin** and **myosin**, rebuild the muscle fibers, making them thicker and stronger.

Protein synthesis is regulated by several factors, including **mechanical tension**, **nutrient availability**, and **hormonal signals**. For muscle growth to occur, the rate of protein synthesis must exceed the rate of protein breakdown. This is known as **positive protein balance**, which is essential for increasing muscle mass.

Several factors influence protein synthesis, but one of the most critical is the **mTOR pathway** (mechanistic target of rapamycin). The mTOR pathway is activated by mechanical stress (like resistance training), amino acids (particularly leucine), and growth factors. Once activated, mTOR promotes the synthesis of proteins necessary for muscle growth.

In addition to mTOR, **insulin** and **growth hormone (GH)** also play vital roles in regulating protein synthesis. Insulin helps shuttle amino acids into muscle cells, while GH promotes the production of insulin-like growth factor 1 (IGF-1), which is a key mediator of muscle growth.

Hypertrophy and Muscle Repair

Muscle hypertrophy, or the increase in muscle size, occurs when the body adapts to the stress placed on muscles. This is typically the result of consistent **resistance training** or **eccentric exercises**, which involve lengthening the muscle while under tension. This type of training creates small tears, or **microtears**, in the muscle fibers, triggering the body's repair processes.

During recovery, the body responds to these microtears by activating muscle stem cells known as **satellite cells**. These cells proliferate and differentiate into new muscle fibers, fusing with existing fibers to help repair the damage. This process leads to an increase in muscle fiber size, a phenomenon known as **muscle hypertrophy**.

The process of muscle repair and hypertrophy is highly dependent on **nutrition**, particularly **protein intake**. The body needs a steady supply of amino acids, the building blocks of protein, to repair damaged muscle fibers. Additionally, **carbohydrates** help replenish glycogen stores that fuel muscle contractions, and **fats** support hormone production, including the synthesis of growth hormones.

Microtears and Muscle Recovery

Muscle recovery is just as important as the exercise itself when it comes to muscle growth. As mentioned, mechanical stress (from exercise) leads to microtears in muscle fibers, which are essential for the growth process. These tears, though small, create the necessary environment for muscle fibers to regenerate and grow. However, without sufficient recovery time, muscle repair cannot occur efficiently, hindering growth.

Rest and **sleep** are crucial components of the recovery process. During deep sleep, the body's production of **growth hormone** increases, stimulating protein synthesis and muscle recovery. The optimal recovery period varies based on the individual, but it is generally recommended that muscles be given 48 hours of rest between intense workouts to allow for adequate repair.

In addition to natural recovery, several external factors can accelerate muscle regeneration. These include adequate **hydration**, **nutrient timing**, and the use of therapies like **electrical stimulation** and **radiofrequency (RF) therapy**. These therapies can accelerate healing by increasing blood flow to the muscles, enhancing nutrient delivery, and promoting cellular repair processes.

External Stimuli for Muscle Regeneration

While the body has its own mechanisms for muscle growth, external stimuli such as **electromagnetic fields**, **RF energy**, and **electrical impulses** can enhance muscle regeneration. Technologies like **RF therapy** and **electromagnetic stimulation (EMS)** provide additional support to the body's natural repair processes by stimulating the muscles directly.

Radiofrequency (RF) therapy uses electromagnetic waves to penetrate deep into tissues, increasing local temperature and promoting **blood circulation**. This increased circulation helps deliver more nutrients and oxygen to the muscles, speeding up recovery. RF therapy also stimulates the production of **collagen** and other extracellular matrix proteins, which are essential for tissue repair.

Electromagnetic stimulation (EMS), on the other hand, uses electrical impulses to stimulate muscle contractions. EMS mimics the signals sent by the nervous system during voluntary muscle contractions, allowing muscles to contract and relax without physical exertion. This is especially useful for enhancing **muscle strength** and **rehabilitation** after an injury, as it can be used to stimulate muscles even when the individual is not actively training.

Both RF therapy and EMS are becoming increasingly popular in clinical and sports settings as effective tools for enhancing muscle recovery and regeneration, particularly in combination with resistance training and proper nutrition.

The Need for Advanced Technologies in Muscle Growth

As we continue to push the boundaries of muscle science, integrating advanced technologies such as AI-driven muscle growth systems, **RF therapy**, and **electromagnetic stimulation** becomes increasingly important. These technologies can enhance the body's natural muscle-building processes, offering athletes, bodybuilders, and individuals undergoing rehabilitation the ability to optimize their results.

By combining the science of muscle physiology with innovative technologies, we can unlock new levels of performance and recovery. AI can help personalize muscle growth plans, optimizing training schedules and recovery periods based on individual needs. Meanwhile, RF and EMS therapies can complement these efforts by accelerating muscle repair, improving nutrient delivery, and enhancing the overall recovery process.

In the chapters ahead, we will delve deeper into how these technologies are being applied in the realm of muscle regeneration and growth. From stimulating growth hormone release to designing personalized treatment protocols using AI, the future of muscle physiology is poised to undergo a revolution.

This chapter provides a foundational understanding of muscle physiology and growth, setting the stage for exploring how advanced technologies like **RF therapy** and **electromagnetic stimulation (EMS)** can complement the body's natural processes for muscle regeneration. Future chapters will expand on how these external technologies, combined with personalized AI-driven approaches, can optimize muscle health and performance.

Chapter 3: Understanding Growth Hormone

Growth hormone (GH), often referred to as **somatotropin**, plays a pivotal role in muscle growth, fat metabolism, and overall tissue regeneration. In this chapter, we will explore what growth hormone is, how it is produced, and how it drives muscle development. We will also investigate the signaling pathways involved in its release and how its anabolic effects can be harnessed through external stimuli, such as **radiofrequency (RF) therapy**, **electromagnetic stimulation (EMS)**, and **artificial intelligence (AI)** optimization.

What Is Growth Hormone?

Growth hormone is a peptide hormone produced by the **pituitary gland**, a small organ located at the base of the brain. It is crucial for growth, particularly during childhood and adolescence, but it also plays an essential role in adult health by regulating various metabolic processes. GH is responsible for stimulating growth and regeneration in almost every tissue of the body, but its most noticeable effects are on **muscle tissue** and **bone**.

GH's primary role is to regulate the body's **anabolic processes** — those that promote the buildup of muscle, bone, and other tissues. It helps the body use proteins, fats, and carbohydrates more effectively by stimulating the production of proteins and the breakdown of fats. Additionally, it promotes the regeneration of tissue after injury and supports cellular repair, which is critical for muscle recovery and growth.

Growth hormone's effects are largely mediated through the production of **insulin-like growth factor 1 (IGF-1)**, a peptide hormone produced primarily by the liver in response to GH stimulation. IGF-1 is considered the most powerful driver of muscle growth, as it stimulates **muscle cell proliferation** (growth) and **differentiation** (the process by which muscle stem cells become mature muscle fibers). Therefore, growth hormone directly impacts muscle regeneration, hypertrophy, and strength.

The Endocrine System: How Growth Hormone Is Released

The secretion of growth hormone is regulated by the **hypothalamus** in the brain, which sends signals to the pituitary gland to either release or inhibit GH production. This process is complex, involving both stimulatory and inhibitory factors. The main stimulatory factor is **growth hormone-releasing hormone (GHRH)**, which signals the pituitary to release GH into the bloodstream. On the other hand, **somatostatin**, another hormone produced by the hypothalamus, inhibits GH release, maintaining a balance between stimulation and inhibition.

Once released into the bloodstream, growth hormone circulates throughout the body, binding to specific receptors on target cells. It stimulates the liver and other tissues to produce IGF-1, which then acts in an autocrine or paracrine manner to regulate tissue growth and repair. Growth hormone levels fluctuate throughout the day and follow a **circadian rhythm**, typically being highest during deep sleep and after exercise.

Interestingly, factors like **nutrition**, **exercise**, and **sleep** can significantly influence GH release. For example, exercise – particularly **resistance training** – is one of the most effective stimuli for GH secretion. It is believed that **high-intensity exercise** causes mechanical stress on muscle fibers, triggering a release of GH to promote tissue repair and muscle growth.

The Anabolic Process: How Growth Hormone Triggers Muscle Synthesis

The anabolic effects of growth hormone are profound and multi-faceted. Once GH is released and IGF-1 is produced, a cascade of events is set in motion that directly impacts muscle growth and regeneration. These include:

1. **Protein Synthesis**: Growth hormone and IGF-1 stimulate the synthesis of proteins in muscle cells, which is crucial for the repair and growth of muscle fibers. This process is initiated by activating the **mTOR pathway**, a key regulator of cellular growth and protein synthesis. When mTOR is activated, it directs the production of proteins like **actin** and **myosin**, the contractile proteins in muscle fibers.

2. **Cellular Differentiation and Hyperplasia**: In addition to protein synthesis, GH and IGF-1 promote the differentiation of muscle stem cells (called **satellite cells**) into mature muscle fibers. This process not only increases the size of existing muscle fibers but also contributes to muscle fiber regeneration and growth, leading to an increase in muscle mass.

3. **Fat Metabolism**: Growth hormone is also crucial for fat metabolism, as it stimulates the breakdown of **triglycerides** in adipose tissue. This process, called **lipolysis**, helps reduce fat stores, leading to a leaner physique. While muscle growth is the primary anabolic effect, fat metabolism also plays a role in the overall health and appearance of the body.

4. **Bone Density and Strength**: Growth hormone's effects extend beyond muscle tissue to bone tissue. By stimulating the production of IGF-1, GH promotes the growth of bone cells and increases **bone density**. This makes GH a critical component in maintaining skeletal health and strength.

5. **Tissue Regeneration**: Beyond muscles and bones, growth hormone plays a critical role in tissue regeneration after injury. It enhances the body's ability to heal damaged tissues by promoting the proliferation of cells involved in the healing process, including **fibroblasts** (which produce collagen) and **endothelial cells** (which form blood vessels).

The combined effects of these processes contribute to improved strength, muscle mass, and overall recovery. However, it is important to note that growth hormone alone is not sufficient for significant muscle gains — the presence of mechanical stress from exercise, adequate nutrition, and optimal rest is essential to fully harness GH's potential for muscle regeneration.

Harnessing Growth Hormone with External Stimuli

Although the body produces growth hormone naturally, various external stimuli can amplify its release and improve muscle regeneration. Technologies like **radiofrequency (RF) therapy** and **electromagnetic stimulation (EMS)**, combined with **AI-driven optimization**, offer innovative methods for boosting growth hormone levels and stimulating anabolic processes.

1. **Radiofrequency (RF) Therapy**: RF energy has been shown to increase blood flow, improve tissue oxygenation, and stimulate collagen production, all of which create an optimal environment for muscle regeneration. Additionally, RF therapy can stimulate the release of growth hormone, further enhancing the recovery and growth of muscle tissue. By increasing local blood circulation, RF therapy aids in the delivery of nutrients, such as amino acids and glucose, that are essential for muscle repair.

2. **Electromagnetic Stimulation (EMS)**: EMS uses electrical impulses to mimic the action potential signals that would normally be sent by the nervous system during physical activity. This stimulates the muscles to contract, resulting in increased blood flow and muscle activation. By modulating the intensity and frequency of these electrical signals, EMS can also stimulate growth hormone release, promoting muscle growth even when the individual is not actively training.

3. **AI-Driven Optimization**: Artificial intelligence plays a key role in optimizing the use of RF and EMS therapies. By tracking real-time data, AI can adjust treatment parameters to maximize growth hormone release and muscle regeneration. Through data collection and analysis, AI can also create personalized treatment plans, ensuring that individuals receive the optimal level of stimulus for their specific physiological needs.

The integration of growth hormone-stimulating technologies with muscle growth strategies creates a powerful synergy. When external technologies like RF and EMS are used in conjunction with natural physiological processes, the body's ability to repair, regenerate, and grow muscle tissue is significantly enhanced. This synergy allows individuals to experience faster recovery, greater muscle gains, and improved overall performance.

In the next chapters, we will explore how these technologies can be tailored for individual needs through AI-driven systems, enhancing the process of muscle growth and recovery. By understanding how growth hormone operates at the cellular level and how it can be optimized through innovative therapies, we lay the groundwork for a comprehensive approach to muscle regeneration that goes beyond traditional training methods.

This chapter provides a comprehensive understanding of **growth hormone's role** in muscle growth, regeneration, and fat metabolism. By exploring its anabolic effects and how external stimuli such as RF therapy, EMS, and AI optimization can enhance GH secretion, we have laid the foundation for utilizing these technologies in the pursuit of optimal muscle health and recovery. In the following chapters, we will explore how to harness the power of AI and other technologies to optimize muscle growth and achieve maximum recovery.

Chapter 4: The Role of AI in Enhancing Muscle Growth

Artificial Intelligence (AI) has evolved from a tool of data analysis to a transformative force that can optimize and enhance muscle growth, recovery, and performance. In the context of muscle regeneration, AI is beginning to play a crucial role in personalizing training, optimizing recovery, and combining innovative technologies like **radiofrequency (RF) therapy** and **electromagnetic stimulation (EMS)** for maximum benefit. This chapter explores how AI contributes to the muscle-building process by leveraging data, personalizing treatments, and facilitating biofeedback mechanisms.

Artificial Intelligence and Data

The core of AI's effectiveness in muscle growth is its ability to handle and analyze large datasets in ways that humans cannot. AI algorithms are trained to process data from various sources, such as **wearables**, **fitness apps**, and **biofeedback devices**, and extract meaningful insights that can guide muscle-building strategies. For example, AI can track your daily workout routine, heart rate variability, sleep patterns, nutrition intake, and muscle performance, compiling all of this information into a comprehensive profile.

This data is then analyzed using machine learning algorithms, which learn from patterns in your performance, recovery, and physiological responses. Based on this analysis, AI can identify **optimal training loads**, **timing**, and **recovery protocols** that would maximize muscle hypertrophy and regeneration for each individual.

One of the most powerful aspects of AI in muscle growth is its ability to continually adapt to the changing needs of the body. For instance, as your muscles grow stronger or your recovery times change, the AI system adjusts your training intensity, rest periods, and nutrition recommendations to reflect these changes. This dynamic optimization helps ensure that you are always pushing your body toward its next level of strength without overtraining or undertraining.

AI in Personalizing Treatment

Not all muscles grow at the same rate, nor do they respond to training in the same way. Factors such as genetics, diet, age, and even sleep habits can influence how well an individual's muscles will grow in response to exercise. This is where AI's power of **personalization** becomes invaluable.

Using advanced **data analytics** and AI algorithms, personalized muscle growth plans can be created. These plans are based on the specific physiological profile of the individual, which includes their unique muscle fiber composition, metabolism, and genetic predispositions. AI takes all of this into account to create a **tailored regimen** that maximizes results while reducing the risk of injury or overtraining.

For instance, if an individual is predisposed to fast-twitch muscle fibers, AI might recommend a combination of high-intensity interval training (HIIT) and explosive weightlifting techniques. Conversely, for someone with a higher proportion of slow-twitch fibers, a more endurance-focused training approach might be optimal, emphasizing longer, moderate-intensity workouts.

Additionally, AI-driven systems can adjust other elements of the muscle growth process, such as **nutritional strategies**. By analyzing data on nutrient intake and muscle performance, AI can recommend a specific ratio of macronutrients (carbohydrates, proteins, and fats) that would best support muscle recovery and growth for that individual. For example, it might suggest increasing protein intake during recovery phases or adjusting carbohydrate levels around workouts to optimize energy use.

AI-Driven Biofeedback

One of the most advanced applications of AI in muscle growth is its ability to provide **real-time biofeedback**. Biofeedback involves using real-time data to adjust muscle training, recovery, and nutrition plans dynamically. Through sensors embedded in wearables or specialized equipment, AI can measure variables such as muscle activation, heart rate, and even electrical activity in the muscle fibers.

For instance, **electromyography (EMG)** sensors can monitor the electrical activity of muscles during workouts, providing valuable information about muscle engagement. AI can interpret this data and give immediate feedback on whether a muscle is being activated optimally during a specific exercise. If the AI detects that the muscle is not fully engaging, it may suggest adjustments to form, the weight being lifted, or the exercise itself to ensure that the muscle is working as efficiently as possible.

Additionally, AI-driven biofeedback can be used to monitor **muscle fatigue levels** during a workout. As muscles approach fatigue, the system can recommend shorter rest periods, changes in exercise intensity, or adjustments to the exercise order to optimize recovery and performance. By continuously monitoring muscle response, the AI ensures that muscles are being pushed to their limits in a safe and controlled manner, preventing overtraining or undertraining.

Beyond exercise, AI can also monitor recovery through real-time feedback on parameters like **heart rate variability (HRV)** and **sleep quality**. For example, if the AI detects that an individual's HRV is significantly lower than baseline levels, it can recommend additional rest or a change in the workout schedule to allow for better recovery.

In some advanced systems, **wearable devices** provide biofeedback during sleep as well. Since muscle repair and growth predominantly occur during rest, AI can track sleep patterns and quality, providing recommendations to enhance sleep hygiene, optimize rest cycles, and improve the recovery process.

AI in Recovery Optimization

Recovery is a critical phase of the muscle-building process, and AI can play a vital role in enhancing recovery strategies. With the wealth of data collected during workouts and recovery periods, AI can develop highly specific recovery protocols. These protocols may include recommendations for **active rest days**, **recovery-focused nutrition**, **sleep optimization**, and the use of complementary therapies like **RF therapy** and **EMS**.

For example, AI may analyze the amount of muscle soreness reported after a workout and recommend a targeted use of **electromagnetic stimulation** or **RF therapy** to accelerate muscle healing. These therapies enhance circulation, promote collagen production, and stimulate muscle fibers to speed up recovery.

AI can also calculate the **optimal rest period** based on the intensity and volume of the workout, ensuring that muscles are given the proper time to repair and regenerate. Furthermore, AI can suggest adjustments in hydration levels, electrolyte balance, and macronutrient distribution to provide the muscles with the exact resources needed for repair.

The Future of AI in Muscle Growth

As AI technology continues to advance, its applications in muscle growth and regeneration will only become more sophisticated. Future AI systems will likely incorporate even more real-time data from a variety of sources, such as genetic information, biomechanical analysis, and hormonal feedback, to create a highly individualized and comprehensive muscle growth plan.

Moreover, AI integration with other cutting-edge technologies like **wearable exoskeletons**, **neurostimulation devices**, and **gene therapy** could enable even greater control over muscle performance and growth. These innovations would provide athletes, fitness enthusiasts, and individuals undergoing rehabilitation with precise, real-time insights into their muscle health, allowing them to fine-tune their approach to muscle growth in ways that were once unimaginable.

The potential for AI to reshape the way we approach fitness, health, and performance is immense. From enhancing **workout efficacy** to accelerating **recovery** and even **preventing injuries**, AI is poised to play an integral role in the future of muscle growth and regeneration.

Conclusion

AI's role in muscle growth and recovery is already transformative, and as technology continues to evolve, its impact will grow exponentially. By personalizing treatment plans, offering real-time biofeedback, and optimizing recovery strategies, AI can help individuals achieve their muscle-building goals more efficiently and safely. The synergy between AI, **RF therapy**, **EMS**, and other technologies promises a future where muscle growth is not just a natural process but one that can be strategically optimized for maximum benefit. In the next chapters, we will explore how AI can be combined with RF and EMS to create a powerful system for muscle regeneration and growth.

Chapter 5: Radiofrequency (RF) Therapy Explained

Radiofrequency (RF) therapy is rapidly becoming one of the most prominent non-invasive treatments used for muscle regeneration, recovery, and overall wellness. By harnessing electromagnetic energy, RF therapy can stimulate deep tissues, improve circulation, promote collagen production, and accelerate healing processes. In this chapter, we will explore the science behind RF therapy, its mechanism of action, and its specific role in muscle growth and recovery. We will also look at real-world case studies where RF therapy has been successfully integrated into muscle regeneration protocols.

What Is RF Therapy?

Radiofrequency therapy involves the application of **radiofrequency waves** to the body's tissues. These waves are a form of electromagnetic energy that oscillate at a specific frequency. The energy produced by RF therapy penetrates deep into the skin and underlying tissues, where it produces thermal and non-thermal effects. RF therapy is commonly used for **skin rejuvenation**, **pain management**, and **muscle regeneration**.

RF waves have the ability to penetrate several layers of the skin, reaching tissues such as muscles, ligaments, and even bones. This makes RF therapy particularly effective in **muscle regeneration** and recovery, where deeper tissues need stimulation for healing and growth. Unlike superficial treatments, RF therapy's ability to target deeper tissues helps promote more comprehensive healing.

Mechanism of Action

RF therapy works through **thermal effects** and **non-thermal effects**, which combine to create an environment conducive to muscle recovery, collagen production, and tissue repair.

- **Thermal Effects**: When RF energy is applied to the body, it generates heat in the tissues. This increase in temperature promotes **vasodilation** (widening of blood vessels), which enhances blood flow and oxygen delivery to the treated area. Increased circulation helps accelerate the removal of toxins and metabolic waste products, speeding up recovery and reducing inflammation. Additionally, the heat from RF therapy stimulates the production of **collagen**, a protein that is essential for tissue repair and muscle regeneration.
- **Non-Thermal Effects**: RF energy also has non-thermal effects that help stimulate the production of **growth factors** and **cellular repair**. These effects occur at the cellular level, where RF waves influence cellular activity, promoting the regeneration of damaged tissues. **Fibroblasts**, the cells responsible for producing collagen, are activated by RF energy, leading to stronger and more resilient tissue formation.

RF Therapy for Muscle Regeneration

RF therapy has shown promising results in muscle regeneration and recovery, particularly in the context of **muscle strain** and **injury recovery**. The combination of **improved circulation, collagen production**, and **enhanced tissue repair** makes RF therapy an ideal treatment modality for individuals seeking faster recovery from muscle fatigue or injury.

RF therapy can also be used to enhance the effectiveness of strength training and muscle-building efforts. By stimulating the tissues, improving blood flow, and encouraging collagen production, RF therapy helps create an optimal environment for muscle growth and healing, even when the body is at rest.

For instance, RF therapy is used by athletes to accelerate muscle recovery after high-intensity workouts or competition. By applying RF energy to fatigued muscles, it helps reduce soreness and enhances muscle repair by stimulating the production of collagen and growth factors. This results in faster recovery times and allows athletes to train harder and more frequently, leading to enhanced muscle growth over time.

Case Studies of RF Therapy for Muscle Recovery

- **Case Study 1: Professional Athletes and Muscle Recovery**

  A professional football player recovering from a hamstring injury used RF therapy as part of his rehabilitation program. After just a few sessions of RF treatment, he reported reduced muscle soreness and improved range of motion. Within weeks, he was able to return to full training, with his muscles regenerating faster than expected. This case highlights RF therapy's ability to accelerate healing and support the rehabilitation process.

- **Case Study 2: Muscle Regeneration Post-Training**

  A bodybuilder incorporated RF therapy into his post-workout recovery routine. After his intense weightlifting sessions, he applied RF energy to target the major muscle groups used during training. Over the course of several months, he noticed a significant improvement in his muscle recovery time and an increase in muscle size and strength. The combination of consistent RF therapy and proper nutrition helped him optimize his muscle-building efforts.

- **Case Study 3: RF for Collagen Production in Muscle Tissue**

    A middle-aged woman experiencing the early stages of **sarcopenia** (muscle loss due to aging) used RF therapy to combat muscle degeneration. By targeting areas of muscle weakness, RF therapy helped improve collagen synthesis and muscle repair. She reported feeling stronger and experiencing fewer aches and pains, demonstrating RF's potential in aging populations to mitigate muscle loss and improve overall muscle health.

RF Therapy and Muscle Growth

While RF therapy is often used to enhance recovery, it also plays a key role in **muscle growth**. By improving **blood circulation**, **oxygen delivery**, and stimulating **collagen production**, RF therapy helps create a favorable environment for muscle hypertrophy (growth). These effects are especially important after intense workouts that cause microtears in muscle fibers, as they speed up the repair process, ensuring that muscle fibers regenerate stronger and thicker.

Additionally, **growth factors** like **IGF-1** (Insulin-like Growth Factor 1) and **VEGF** (Vascular Endothelial Growth Factor) are activated during RF treatment, promoting muscle regeneration and vascularization. Increased blood flow and oxygenation not only improve recovery times but also ensure that muscles have the necessary nutrients to grow and strengthen.

Furthermore, RF therapy's ability to stimulate the production of **collagen** strengthens the **extracellular matrix** surrounding muscle fibers. This provides better structural support for growing muscle fibers, enhancing their ability to grow under stress and tension. The result is faster muscle growth, reduced recovery time, and improved muscle quality.

Combining RF Therapy with Other Techniques

For even more effective results, RF therapy can be combined with other muscle regeneration techniques, such as **electromagnetic stimulation (EMS)** and **AI-driven optimization**. By using RF therapy to stimulate deep tissue and enhance circulation, and EMS to directly contract muscles, individuals can create a comprehensive muscle recovery and growth strategy.

In addition, AI can help monitor and adjust RF therapy treatments for the best possible outcome. By analyzing the individual's muscle response to therapy, AI can personalize the RF treatment plan, adjusting parameters such as intensity, frequency, and duration to optimize recovery and muscle regeneration.

Safety and Side Effects

RF therapy is generally safe when performed under the supervision of trained professionals. The procedure is non-invasive, requires no downtime, and has minimal side effects. Some individuals may experience mild redness or warmth in the treated area, but these effects typically subside shortly after treatment.

However, it is important for individuals with certain medical conditions—such as **heart disease**, **skin conditions**, or **metal implants**—to consult with a healthcare provider before undergoing RF therapy. When used correctly, RF therapy offers a safe and effective treatment for muscle growth, recovery, and regeneration.

Conclusion

Radiofrequency (RF) therapy has emerged as a powerful tool for muscle regeneration, recovery, and growth. Its ability to stimulate blood circulation, promote collagen production, and enhance tissue repair makes it an invaluable part of any muscle-building or rehabilitation regimen. As we continue to integrate advanced technologies like **electromagnetic stimulation (EMS)** and **AI-driven optimization**, RF therapy will play an increasingly important role in enhancing muscle growth and improving recovery outcomes. Whether for athletes, bodybuilders, or individuals recovering from injury, RF therapy offers a non-invasive, highly effective solution for accelerating muscle regeneration and promoting long-term muscle health.

Chapter 6: Electromagnetic Stimulation (EMS) and Its Applications

Electromagnetic Stimulation (EMS) is a powerful therapeutic tool that has gained significant attention in recent years for its ability to enhance muscle growth, promote recovery, and aid in rehabilitation. EMS utilizes electrical impulses to stimulate muscles, mimicking the natural nerve signals that trigger muscle contractions. By understanding how EMS works, its applications in muscle growth and recovery, and the ways in which it is used in clinical practice, we can appreciate its profound impact on optimizing muscle performance.

What Is EMS?

Electromagnetic Stimulation (EMS) is a technique that uses electrical impulses to induce muscle contractions. The electrical impulses are transmitted through electrodes placed on the skin, and these signals mimic the action potentials that the brain sends to muscles during voluntary movement. EMS devices can generate a range of frequencies and intensities, enabling them to target various muscle fibers and tissues.

In the context of muscle growth and recovery, EMS is particularly valuable because it provides a way to stimulate muscles without the need for voluntary effort. This means that muscles can be activated even when the person is at rest, making EMS a useful tool in both rehabilitation and performance enhancement.

EMS works by delivering controlled electrical impulses through the skin via electrodes placed over the muscles. The impulses then travel through the skin to the underlying muscle fibers, where they trigger a **muscle contraction**. The type, frequency, and duration of the electrical impulses can be adjusted to target specific muscle groups and achieve various therapeutic goals, such as **muscle strengthening**, **pain relief**, or **rehabilitation**.

How EMS Works on Muscles

The primary mechanism behind EMS is the stimulation of **motor neurons**, which are responsible for activating muscle fibers. When the electrical impulse reaches the motor neurons, they transmit the signal to the muscle fibers, causing them to contract. These contractions are similar to the ones that occur during physical activity, but they are generated externally through the EMS device rather than by voluntary muscle activation.

The contractions induced by EMS can be categorized into two main types:

- **Isometric Contractions**: These occur when the muscle contracts but does not change length. Isometric contractions are particularly useful for strengthening muscles without straining joints or tissues.
- **Isotonic Contractions**: These involve both the shortening and lengthening of the muscle, mimicking the actions involved in exercise. This type of contraction is effective in enhancing muscle endurance and strength.

By adjusting the **frequency** and **intensity** of the electrical impulses, EMS can target specific muscle fibers, such as **fast-twitch** fibers (which are responsible for explosive movements) or **slow-twitch** fibers (which are used for endurance activities). This adaptability makes EMS a versatile tool in both muscle-building and rehabilitation programs.

EMS for Injury Recovery and Rehabilitation

EMS has been widely used in clinical settings, particularly for **injury recovery** and **rehabilitation**. It is especially valuable for individuals who are unable to engage in traditional exercise due to injuries, surgeries, or other physical limitations. EMS helps stimulate muscles and improve circulation, aiding in the recovery process and preventing muscle atrophy (muscle wasting).

In rehabilitation, EMS is often used to treat conditions such as:

- **Muscle Atrophy**: When a muscle is not used due to an injury or surgery, it can begin to shrink and lose strength. EMS can help maintain muscle size and strength during recovery by stimulating the muscles and promoting blood flow.
- **Muscle Spasms and Cramps**: EMS can help relieve muscle spasms and cramps by promoting relaxation and improving circulation to the affected area.
- **Post-Operative Recovery**: After surgeries such as joint replacements or ligament repairs, EMS can be used to accelerate recovery and help restore muscle strength.
- **Neuromuscular Rehabilitation**: For patients who have experienced nerve damage, EMS can stimulate motor neurons and help retrain muscles to respond to nerve signals, promoting functional recovery.

A key benefit of EMS in rehabilitation is its ability to activate muscles without the need for strenuous physical activity. This makes it ideal for people who may be unable to perform regular exercise due to pain or mobility limitations. By inducing muscle contractions in a controlled and safe manner, EMS promotes **muscle healing**, **strengthening**, and **rehabilitation**.

EMS for Muscle Strengthening and Performance Enhancement

While EMS is often associated with rehabilitation, it is also a valuable tool for **muscle strengthening** and **performance enhancement** in athletes and fitness enthusiasts. EMS can be used to **augment traditional training**, improve muscle endurance, and increase strength.

For athletes, EMS can provide the following benefits:

- **Enhanced Muscle Activation**: EMS can target muscle fibers that may not be fully activated during regular exercise, increasing overall muscle recruitment and enhancing training effectiveness.
- **Increased Muscle Endurance**: By stimulating muscles to contract repeatedly, EMS can help improve muscle endurance and resistance to fatigue.
- **Improved Recovery**: EMS is used by athletes post-training to speed up recovery, reduce soreness, and improve blood flow to muscles, facilitating quicker muscle regeneration.
- **Complementary Training Tool**: EMS can be used in conjunction with traditional resistance training to provide additional stimulation and promote greater muscle hypertrophy (growth).

Research has shown that **neuromuscular electrical stimulation** can enhance performance by complementing regular exercise routines. EMS devices can be programmed to stimulate muscles during specific phases of the training cycle, making them a valuable tool for **sports performance optimization**.

EMS for Targeted Muscle Development

EMS offers a unique advantage in **targeted muscle development**. Unlike traditional workouts, which may engage multiple muscle groups, EMS allows for **isolated muscle activation**. This can be particularly useful for individuals looking to develop a specific muscle group or address muscle imbalances.

For instance, if a person has a weakness or imbalance in a particular muscle, such as the hamstrings or quadriceps, EMS can be used to specifically target that muscle, ensuring it is strengthened and developed. This targeted approach can help athletes and individuals undergoing rehabilitation to address weaknesses and optimize their performance.

EMS is also useful for **toning muscles** in areas that are difficult to isolate with traditional workouts, such as the lower back, abs, or glutes. By using different stimulation patterns, EMS can improve **muscle definition** and **muscle tone**, enhancing overall body composition.

EMS Safety and Precautions

While EMS is generally considered safe, it is important to use the technology properly to avoid injury. Users should always follow the manufacturer's guidelines and ensure that the electrodes are placed correctly on the skin to avoid skin irritation or burns. Additionally, EMS should not be used on certain areas of the body, such as the head, neck, or chest, unless prescribed by a healthcare professional.

Individuals with specific health conditions, such as **heart disease**, **epilepsy**, or **pregnancy**, should consult with a healthcare provider before using EMS. It is also important to note that EMS should not be used as a substitute for regular physical activity. It is most effective when used in conjunction with a balanced exercise program.

Combining EMS with Other Therapies

For optimal results, EMS can be combined with other therapies, such as **RF therapy**, **AI-driven optimization**, and **nutritional support**. By integrating these therapies, individuals can create a comprehensive approach to muscle growth, recovery, and rehabilitation. For example:

- **Combining EMS with RF Therapy**: RF therapy can enhance blood flow and stimulate collagen production, while EMS can directly stimulate muscle contractions. Together, they provide a synergistic effect that accelerates muscle recovery and growth.
- **AI-Driven Muscle Protocols**: AI can help optimize EMS treatments by adjusting the intensity, frequency, and duration based on the individual's progress and needs, ensuring that the treatments are always aligned with their muscle-building goals.

Conclusion

Electromagnetic Stimulation (EMS) is a versatile and effective tool for muscle growth, recovery, and rehabilitation. Whether used for injury recovery, muscle strengthening, or enhancing performance, EMS offers a unique and non-invasive approach to optimizing muscle health. By stimulating muscle contractions, improving circulation, and supporting muscle regeneration, EMS provides a powerful complement to traditional exercise regimens and therapeutic techniques. As technology continues to advance, EMS is poised to play an even more significant role in the future of muscle development and recovery.

Chapter 7: Combining AI with RF and EMS

In the pursuit of enhanced muscle growth, recovery, and regeneration, the integration of **Artificial Intelligence (AI)** with **Radiofrequency (RF) therapy** and **Electromagnetic Stimulation (EMS)** represents a powerful combination that optimizes muscle health. Each of these technologies offers distinct advantages, but when combined, they create a synergistic effect that can significantly accelerate muscle regeneration, improve recovery time, and promote sustained growth. This chapter explores how AI can enhance the efficacy of RF therapy and EMS, offering a more personalized, data-driven, and dynamic approach to muscle growth and rehabilitation.

The Synergy of AI and Electrotherapy

The combination of AI with RF and EMS technologies is an exciting frontier in muscle therapy. By leveraging AI to optimize the parameters of RF and EMS treatments, practitioners can ensure that each therapy session is tailored to the individual's specific needs, physiology, and progress. This integration enhances both the immediate effects of the treatments and their long-term benefits.

RF Therapy uses electromagnetic energy to stimulate deep tissues, improve blood circulation, and promote collagen production. **EMS**, on the other hand, utilizes electrical impulses to induce muscle contractions, targeting specific muscle fibers to enhance strength, recovery, and hypertrophy. AI's role in this synergy is to analyze real-time data from these therapies and provide personalized adjustments that optimize results.

The integration of AI ensures that the **parameters** of RF and EMS—such as **intensity**, **frequency**, and **duration**—are fine-tuned to each individual's muscle recovery needs and goals. For example, AI systems can adapt RF or EMS treatment settings based on **muscle response**, **fatigue levels**, and **progressive adaptation** over time. This dynamic, real-time optimization maximizes the therapeutic impact of both RF and EMS, reducing the risk of overtraining or ineffective treatment.

Real-Time Monitoring and Adjustments

One of the most significant benefits of combining AI with RF and EMS is the ability to monitor treatment outcomes in real time. AI systems can process vast amounts of data from various sources—**wearable sensors**, **muscle activity**, **biometric data**, and **feedback loops**—to provide continuous feedback on the muscle's performance and recovery.

For instance, wearable devices equipped with **electromyography (EMG)** sensors can measure the electrical activity in the muscles during an EMS session. AI can analyze this data to determine whether the muscle is being properly activated and if the treatment is achieving the desired intensity. If the AI detects suboptimal activation, it can adjust the EMS parameters to ensure the muscle is fully engaged, thereby maximizing the therapeutic effects.

Similarly, when applying RF therapy, AI can monitor the skin's temperature and tissue responses, adjusting the RF wave intensity and duration accordingly. By continuously adjusting the RF therapy based on the individual's feedback, AI ensures that the treatment is always effective without causing discomfort or overstimulation.

The ability to make **real-time adjustments** based on continuous monitoring allows for **adaptive training regimens** that evolve as the muscle grows stronger. This kind of personalized therapy helps individuals achieve optimal results while minimizing the risk of injury or overtraining.

Creating Custom Protocols

A key advantage of AI-driven RF and EMS therapy is the creation of **customized treatment protocols** for each individual. Traditional muscle growth programs tend to use one-size-fits-all approaches, which can sometimes lead to inefficient training and recovery. By integrating AI into the process, muscle growth and recovery protocols can be dynamically adapted to meet the specific needs and goals of each person.

AI can generate personalized training schedules based on several factors, including:

- **Muscle Fiber Type**: Different muscle fibers respond to different types of stimulation. AI can create protocols that target fast-twitch fibers for explosive power or slow-twitch fibers for endurance training.
- **Recovery Rates**: AI can analyze data such as **heart rate variability**, **muscle soreness**, and **sleep quality** to determine how much rest an individual requires between sessions. By optimizing recovery times, AI ensures that muscles are not overtrained and have the necessary time to repair and grow.
- **Strength Levels**: By tracking strength gains and performance metrics, AI can adjust the intensity and volume of both RF and EMS treatments to align with the individual's progress.

These custom protocols allow individuals to **target specific muscle groups**, address weak spots, and promote balanced muscle growth. For example, AI might recommend a more intense EMS protocol for the quadriceps while using a milder RF treatment for the hamstrings, based on previous performance and recovery data.

AI-Driven Progress Tracking

Another powerful aspect of integrating AI with RF and EMS is the ability to track **progress over time**. AI systems can gather data from each session, including **muscle activity**, **treatment intensity**, and **recovery metrics**, and use this information to assess the effectiveness of the treatment and identify areas for improvement.

Through this progress tracking, AI can identify trends in muscle growth, recovery time, and strength improvements. It can highlight which treatment protocols are most effective and adjust future sessions accordingly. Additionally, by comparing an individual's data with **population benchmarks**, AI can identify whether the current protocol is too aggressive or too conservative, and recommend adjustments to ensure maximum benefit.

Over time, this data-driven approach helps individuals and practitioners avoid stagnation. AI helps break through **plateaus** by constantly adjusting the training regimen to push muscles beyond their previous limits. The ability to track and adjust progress in real time also encourages a sense of accountability and motivation, as users can directly see the effects of their training protocols.

Integration of AI, RF, and EMS for Comprehensive Muscle Therapy

The synergy between AI, RF therapy, and EMS allows for a comprehensive approach to muscle growth and recovery. This integrated system can support the full cycle of muscle development, from training to recovery, with the precision and personalization needed to optimize results.

- **Pre-Training Optimization**: AI can assess an individual's readiness for exercise by evaluating sleep patterns, muscle soreness, and past performance. It can recommend specific RF or EMS treatments to prepare muscles for intense activity, reducing the risk of injury and enhancing performance.
- **In-Training Enhancement**: During workouts, AI can guide individuals through optimal intensity levels for EMS to ensure that muscles are contracting properly. It can also adjust RF parameters to reduce inflammation and enhance circulation in real time.
- **Post-Training Recovery**: After training, AI can customize recovery protocols based on muscle fatigue and progress. It can recommend the ideal combination of RF therapy and EMS to promote muscle repair, reduce soreness, and stimulate muscle growth.

This comprehensive, AI-powered system provides individuals with a holistic solution for achieving their muscle growth and recovery goals. By integrating AI's data-driven capabilities with the healing power of RF and EMS, individuals can accelerate muscle regeneration, optimize training, and achieve faster, more sustainable results.

Conclusion

The combination of **Artificial Intelligence (AI)**, **Radiofrequency (RF) therapy**, and **Electromagnetic Stimulation (EMS)** represents a groundbreaking approach to muscle growth and regeneration. Through real-time monitoring, personalized protocols, and continuous progress tracking, AI optimizes both RF and EMS therapies for better results. By using AI to customize and adjust the treatments, this integrated approach maximizes the potential of each therapy, creating a powerful synergy that enhances muscle recovery, growth, and performance. This personalized, data-driven system holds the promise of revolutionizing the way we approach muscle health, offering unparalleled benefits in both rehabilitation and muscle optimization. As we continue to explore these technologies, their combined power will be integral to shaping the future of fitness and rehabilitation.

Chapter 8: The Bioelectric Muscle Approach to Hormonal Stimulation

Muscle growth, recovery, and regeneration are complex processes influenced by a variety of factors. One of the most critical elements in this process is the body's hormonal regulation. Specifically, hormones such as **growth hormone (GH)** and **testosterone** play pivotal roles in triggering **protein synthesis**, promoting **muscle hypertrophy**, and enhancing **cellular repair**. In recent years, advances in **bioelectric muscle therapies**—including **Radiofrequency (RF) therapy** and **Electromagnetic Stimulation (EMS)**—have demonstrated their potential in **stimulating hormone release** and optimizing the hormonal balance necessary for muscle growth. This chapter explores how these therapies can be leveraged to activate hormonal pathways, enhance muscle regeneration, and balance hormonal levels to support optimal muscle health.

Stimulating Growth Hormone Release

Growth hormone (GH) is a key player in the regulation of muscle growth. It is produced by the **pituitary gland** and is responsible for stimulating the production of **insulin-like growth factor 1 (IGF-1)**, a hormone that directly promotes muscle regeneration and repair. GH and IGF-1 act in concert to enhance **protein synthesis**, **increase muscle cell proliferation**, and **repair damaged tissues**. The release of growth hormone is typically triggered by factors such as exercise, sleep, and nutrient intake.

RF therapy and **EMS** offer unique, non-invasive ways to stimulate the release of growth hormone, enhancing muscle regeneration and growth. These bioelectric treatments work by influencing the body's **neuroendocrine system**, which governs hormone release in response to external stimuli.

1. **RF Therapy and Growth Hormone**

   RF therapy, which utilizes electromagnetic energy to penetrate tissues, has been shown to stimulate **blood flow** and **collagen production** at the cellular level. Interestingly, RF energy also activates the body's hormonal pathways, encouraging the release of growth hormone. The increased circulation during RF treatment helps to deliver essential nutrients to tissues and organs, including the pituitary gland, which is responsible for growth hormone secretion. The thermal effects of RF waves may further enhance the pulsatile release of growth hormone, contributing to faster tissue repair and muscle regeneration.

2. **EMS and Hormonal Stimulation**

   EMS has been shown to improve muscle function by inducing **muscle contractions** through electrical impulses. While the primary goal of EMS is to stimulate muscle fibers, research has also suggested that EMS can influence **hormonal regulation**, particularly growth hormone levels. The electrical impulses generated during EMS therapy have been shown to stimulate **neurotransmitter** release, triggering the brain and endocrine system to release **growth hormone**. This effect is particularly beneficial when used after exercise or during recovery phases, as it can accelerate muscle healing by boosting the body's natural anabolic processes.

By combining RF therapy and EMS, it is possible to create a dual stimulation effect that maximizes growth hormone release, ensuring that muscles receive optimal support during recovery and growth.

Hormonal Pathways Activation

The body's hormonal pathways are responsible for regulating a wide range of physiological processes, including metabolism, protein synthesis, and muscle recovery. **Electromagnetic stimulation**, particularly RF therapy and EMS, has been found to activate several key hormonal pathways involved in muscle regeneration.

- **Growth Hormone and IGF-1**: As mentioned, RF and EMS therapies can stimulate the pituitary gland to release growth hormone. This hormone then stimulates the production of IGF-1 in the liver, which in turn promotes muscle growth and repair by activating **satellite cells**—the muscle stem cells responsible for regenerating muscle tissue.
- **Testosterone**: Another hormone involved in muscle growth is **testosterone**, which plays a crucial role in stimulating protein synthesis and enhancing muscle hypertrophy. Research has shown that both RF therapy and EMS can influence **testosterone levels**, especially when used in conjunction with strength training or resistance exercises. By stimulating the release of testosterone, these therapies can enhance muscle growth and strength gains.
- **Cortisol Regulation**: **Cortisol**, the body's primary stress hormone, can have a catabolic effect on muscles when levels remain elevated for extended periods. Both RF therapy and EMS have been shown to help regulate cortisol levels, ensuring that the body remains in an anabolic (muscle-building) state, even during periods of physical stress.

By activating these hormonal pathways, RF and EMS therapies help create an optimal environment for muscle growth, promoting tissue regeneration and enhancing recovery. These therapies are particularly useful in overcoming the **plateaus** that individuals often experience in traditional strength training programs, as they ensure that hormonal pathways remain activated throughout the recovery and regeneration process.

Optimizing Hormonal Balance for Muscle Growth

Consistent stimulation of growth hormone and other key hormones is essential for muscle regeneration and hypertrophy. However, the body's hormonal balance must be carefully managed to optimize muscle growth. Overactivation or dysregulation of certain hormones can lead to adverse effects such as **muscle breakdown**, **fatigue**, or **increased injury risk**.

AI-driven optimization combined with RF and EMS therapies offers a way to precisely control and monitor hormonal balance. AI systems can analyze real-time data from wearables, fitness trackers, and biofeedback devices to monitor changes in muscle performance, recovery rates, and **biomarkers** related to hormonal levels (such as cortisol, testosterone, and growth hormone). Based on this data, AI can adjust the frequency, intensity, and duration of RF and EMS treatments to ensure that hormone levels are **optimized** for muscle growth while avoiding potential imbalances.

- **Personalized Protocols**: AI can tailor hormonal stimulation protocols to each individual's needs. For example, an athlete who is training for high-intensity performance may require more frequent stimulation of growth hormone, while someone recovering from injury may need a protocol designed to enhance tissue repair without overstimulating cortisol production.
- **Monitoring Rest and Recovery**: AI can also monitor **sleep patterns** and **rest cycles**, two critical factors in hormonal release. Since growth hormone is predominantly released during deep sleep, AI can recommend specific times for RF and EMS treatment to align with the body's natural circadian rhythms, optimizing hormonal balance during recovery periods.

By integrating AI-driven hormonal optimization with RF and EMS therapies, individuals can achieve consistent, controlled muscle growth that minimizes the risks associated with excessive hormone levels or prolonged periods of imbalance.

The Future of Hormonal Stimulation in Bioelectric Muscle Therapy

As research continues into the effects of RF, EMS, and AI on muscle growth and hormonal regulation, the potential for these technologies to revolutionize muscle regeneration is immense. The ability to stimulate **growth hormone release** through non-invasive, **personalized treatments** will pave the way for new recovery protocols and muscle-building strategies, especially for individuals in rehabilitation, older adults experiencing **sarcopenia**, or athletes seeking optimized performance.

- **Stem Cell Therapy and Bioelectric Stimulation**: The combination of bioelectric therapies with **stem cell therapy** represents an exciting future direction for hormonal stimulation. Electrical signals have been shown to influence **stem cell behavior**, and combining RF, EMS, and stem cell treatments may allow for even more effective muscle regeneration and healing.
- **Gene Therapy and AI**: With advancements in gene therapy and the use of **CRISPR** technologies, it may be possible to directly influence hormonal pathways to enhance muscle growth. AI systems could potentially be used to track genetic responses to RF and EMS therapies, allowing for the development of treatments that are precisely tailored to an individual's **genetic makeup**.

Conclusion

Bioelectric muscle therapies such as **RF therapy** and **EMS** provide a powerful means of stimulating hormonal release, activating key hormonal pathways, and optimizing the body's anabolic environment. By targeting growth hormone, testosterone, and cortisol regulation, these therapies ensure that the body remains in an ideal state for muscle growth and repair. AI-driven personalization of these therapies enhances their effectiveness, allowing for continuous adaptation based on real-time data.

As we continue to advance our understanding of bioelectric therapies and hormonal stimulation, these technologies will become integral to not only enhancing athletic performance but also improving **muscle regeneration** in clinical and rehabilitation settings. The potential to combine **AI**, **RF**, **EMS**, and **hormonal optimization** opens the door to a future where muscle growth and recovery can be both scientifically and personally tailored for every individual.

Chapter 9: Integrating Nutrition with Bioelectric Stimulation

Muscle growth, recovery, and regeneration depend not only on physical training and therapeutic interventions but also on the vital role that **nutrition** plays in fueling the bioelectric processes within the body. The integration of **AI, Radiofrequency (RF) therapy**, and **Electromagnetic Stimulation (EMS)** with a well-balanced, strategic diet creates a synergistic approach to muscle development. By fueling the body with the right nutrients, we can enhance the bioelectric signals used to stimulate muscle growth, repair, and regeneration. This chapter explores the essential role of nutrition in supporting bioelectric stimulation therapies, how specific nutrients enhance muscle recovery, and how combining these elements can optimize muscle health.

Role of Nutrition in Muscle Growth

Muscle growth occurs when muscle fibers undergo **hypertrophy** (growth) as a response to mechanical stress. This process requires more than just physical exertion—it demands an adequate supply of **nutrients** to support the recovery and rebuilding of muscle tissue.

Key components of an effective muscle-building diet include **proteins, carbohydrates**, and **fats**, each of which serves an important function in the muscle regeneration process:

Proteins

muscle protein synthesis

microtears

Whey protein

casein

plant-based proteins

- **Leucine**: This essential amino acid plays a particularly significant role in initiating muscle protein synthesis. Leucine activates the **mTOR pathway**, a key signaling pathway for muscle growth. This is where **AI-driven muscle protocols** and **RF therapy** can enhance recovery by promoting the release of amino acids at the cellular level.
- **BCAAs** (Branched-Chain Amino Acids): BCAAs, particularly **valine, leucine**, and **isoleucine**, are known to help reduce muscle breakdown during periods of intense physical activity, ensuring that muscle growth is optimized and that recovery is quickened.

Carbohydrates

Glycogen

Insulin Response

AI-enhanced muscle recovery protocols

Fats

hormonal health

testosterone

growth hormone

avocados

olive oil

nuts

fatty fish

omega-3 fatty acids

Testosterone Regulation

testosterone

Supplements and Bioelectric Stimulation

When combined with bioelectric stimulation therapies such as RF and EMS, **supplements** can enhance the effectiveness of muscle regeneration and growth. By boosting the bioelectric signals that RF and EMS therapies use to stimulate muscle fibers, the right supplements can accelerate recovery and improve results.

1. **Creatine**:

   Creatine is a naturally occurring compound found in muscle cells that helps produce energy during high-intensity activities. Supplementing with creatine has been shown to increase **muscle mass**, enhance **strength**, and improve **muscle recovery** by increasing **phosphocreatine** stores in muscles, which helps regenerate ATP (adenosine triphosphate), the body's primary energy molecule.

   When used in combination with **EMS therapy**, creatine can enhance the **muscle contractions** stimulated by EMS impulses, leading to more powerful muscle activation and quicker muscle growth. Similarly, **RF therapy** can help improve **muscle tissue regeneration** by increasing nutrient delivery, further amplifying the benefits of creatine supplementation.

2. **Amino Acids and Peptides**:

   The use of **amino acids** (such as **glutamine**, **arganine**, and **glutathione**) in supplement form can aid in muscle repair by promoting protein synthesis. These amino acids are vital in repairing muscle tissues broken down during intense exercise or muscle injury.

Peptides

BPC-157

TB-500

collagen production

inflammation

Vitamins and Minerals

Vitamin D

muscle strength

AI-optimized RF treatments

magnesium

zinc

testosterone production

Fueling the Bioelectric Process

Nutrition does not only influence muscle recovery after training; it also enhances the **bioelectric signals** that drive muscle regeneration. By providing the body with the right nutrients at the right times, we can optimize the effects of RF and EMS therapies.

- **Timing**: **Nutrient timing** plays an important role in muscle recovery. Consuming **protein** and **carbohydrates** immediately after a workout optimizes the **insulin response**, helping deliver amino acids and glucose to muscles when they are most receptive. Pairing this with **AI-driven adjustments** to RF and EMS therapy ensures that nutrients are delivered efficiently to the muscles, speeding up recovery and promoting muscle growth.
- **Synergistic Effects**: By integrating **bioelectric muscle therapies** with a diet rich in muscle-building nutrients, individuals can create a **synergistic effect** that accelerates muscle regeneration. For example, the anti-inflammatory properties of **omega-3 fatty acids** combined with the **stimulated blood flow** from RF therapy can reduce muscle soreness and accelerate recovery. The same synergy applies to the combination of **branched-chain amino acids (BCAAs)** and **EMS**, which work together to enhance **muscle endurance** and recovery by reducing protein breakdown.

Integrating Nutritional Strategies with Bioelectric Stimulation for Optimal Results

The integration of **nutrition** with **bioelectric stimulation** therapies like RF and EMS creates a comprehensive approach to muscle growth and recovery. Whether used by athletes, bodybuilders, or individuals recovering from injury, this integrated strategy enhances the body's natural muscle-building processes and speeds up recovery.

- **Personalized Nutrition Plans**: AI can monitor real-time feedback from wearables, sensors, and treatment devices to tailor **dietary plans** and **bioelectric treatments** to the individual's specific needs. For example, an individual recovering from a muscle injury may benefit from a higher intake of **glutamine** and **amino acids**, combined with **RF therapy** to reduce inflammation and promote healing.
- **Optimized Recovery Cycles**: By tracking data on sleep quality, nutrient intake, and muscle activity, AI can suggest **rest cycles** that optimize both **muscle recovery** and **hormonal balance**, ensuring that muscle tissues are receiving the right nutrients and therapies at the ideal times for regeneration.

Conclusion

Integrating **nutrition** with **bioelectric muscle stimulation** therapies such as **RF therapy** and **EMS** is key to optimizing muscle growth, recovery, and regeneration. By fueling the body with essential nutrients—such as proteins, carbohydrates, fats, and key micronutrients—individuals can enhance the bioelectric signals that drive muscle growth. Combined with the precision of AI-driven optimization, this integrated approach ensures that both **training** and **recovery** are maximized, promoting muscle health at every stage. As we continue to evolve our understanding of how **nutrition** and **bioelectric therapies** work together, the potential for faster recovery, enhanced performance, and sustained muscle growth becomes even greater.

Chapter 10: RF and EMS in Clinical Practice

The integration of **Radiofrequency (RF) therapy** and **Electromagnetic Stimulation (EMS)** has transformed clinical practices, offering innovative and non-invasive approaches to muscle recovery, pain management, and overall rehabilitation. These bioelectric therapies have shown tremendous potential in a variety of clinical settings, from **sports medicine** to **physiotherapy** and **chronic pain management**. This chapter explores the current applications of RF and EMS in medical practice, providing real-world examples, case studies, and an overview of the safety and efficacy of these treatments.

Current Applications in Medicine

RF and EMS technologies are increasingly being used in **medical treatments** to promote healing, restore muscle function, and accelerate recovery from injuries. These therapies are applied across a variety of clinical disciplines, such as **orthopedics**, **neurology**, and **sports medicine**.

1. **Physical Rehabilitation and Muscle Recovery** RF and EMS are particularly valuable in **rehabilitation settings**, where they help patients regain muscle strength, improve flexibility, and restore mobility after surgery, injury, or extended periods of immobility. The ability of **EMS to stimulate muscle contraction** allows patients who have been immobilized due to surgery or injury to maintain muscle activity and prevent **atrophy**. By using EMS in combination with physical therapy, patients can expedite the recovery process and regain their full range of motion. RF therapy is used to **stimulate deep tissues**, improving **circulation, collagen production**, and **muscle regeneration**. This is particularly useful in the treatment of soft tissue injuries such as **muscle strains, ligament sprains**, and **tendonitis**. RF waves promote **cellular repair** and **reduce inflammation**, allowing for quicker recovery and reducing the risk of chronic injuries.
2. **Chronic Pain Management** One of the most common uses for RF and EMS in clinical practice is in the management of **chronic pain**. Conditions such as **arthritis, sciatica, fibromyalgia**, and **musculoskeletal pain** often benefit from RF and EMS therapies.

- **RF Therapy**: RF waves penetrate deeply into the tissue, providing **analgesic effects** and **reducing inflammation**. In the case of **chronic pain conditions**, RF therapy is often used to treat **deep muscle pain**, **joint pain**, and **neuropathic pain**. By stimulating **nerve cells**, RF therapy helps to **block pain signals** and promote the release of **endorphins**, which are natural painkillers.
- **EMS**: EMS therapy has shown effectiveness in **reducing pain** and **improving muscle function** for patients with chronic pain. By stimulating muscle contractions, EMS can enhance blood circulation and **reduce muscle tightness** and **spasms**. This is particularly beneficial for conditions like **lower back pain** or **neck pain**, where tight muscles contribute to discomfort and limited movement.

Post-Surgical Rehabilitation

accelerate recovery

reduce swelling

promote tissue healing

joint replacement surgery

risk of deep vein thrombosis (DVT)

scar tissue formation

Case Studies: Real-World Applications

1. **Case Study 1: Post-Injury Muscle Recovery in Athletes** A professional **soccer player** who suffered a **hamstring injury** incorporated both **RF therapy** and **EMS** into his rehabilitation program. The goal was to stimulate muscle healing and reduce inflammation. EMS was used daily to maintain muscle tone and prevent atrophy, while RF therapy was applied to improve circulation and promote collagen production for tissue regeneration.

After four weeks of this combined treatment, the athlete reported a significant reduction in pain and swelling, as well as improved range of motion. By incorporating these bioelectric therapies into his recovery program, the athlete was able to return to training more quickly, with enhanced muscle strength and mobility.

2. **Case Study 2: Chronic Pain Management in Older Adults** An **elderly patient** with **osteoarthritis** in the knee sought RF and EMS therapies to manage chronic pain and improve joint mobility. RF therapy was applied to the affected knee area to stimulate **deep tissues**, reduce inflammation, and enhance **collagen production** for joint regeneration. EMS was used to stimulate the surrounding muscles to prevent further muscle loss and maintain joint stability.

Over the course of eight weeks, the patient experienced significant pain relief and improved knee function, allowing for better mobility and reduced reliance on pain medications. The combined application of RF and EMS therapies offered a non-invasive, drug-free solution to managing chronic pain and improving quality of life.

3. **Case Study 3: Post-Surgery Recovery for Joint Replacement** A patient recovering from a **total hip replacement** used a combination of RF therapy and EMS to speed up recovery. RF therapy was used to promote **muscle regeneration**, reduce **inflammation**, and enhance **circulation** in the area surrounding the new joint. EMS therapy helped maintain muscle strength and flexibility during the rehabilitation process, preventing muscle atrophy and promoting joint function.

   After six weeks, the patient was able to walk without assistance, with significantly less pain and swelling. By integrating RF and EMS into the post-surgical rehabilitation program, the patient was able to achieve a faster recovery, with improved muscle strength and joint mobility.

Safety and Efficacy: An Overview

Both **RF therapy** and **EMS** have demonstrated high safety and efficacy in clinical settings when used according to established protocols and under professional supervision. However, like all therapeutic interventions, these treatments do have specific contraindications and safety guidelines that should be followed.

RF Therapy Safety Considerations

- RF therapy is generally considered safe, but it should be avoided in patients with certain conditions, such as **pacemakers**, **implanted defibrillators**, or **metal implants** in the area being treated. The electromagnetic energy may interfere with these devices.
- Pregnant women should also avoid RF treatments, as the effects of RF energy on pregnancy have not been fully studied.

2. To ensure safety, **professional supervision** is crucial when administering RF therapy, particularly in clinical settings. Sessions should be **time-limited** and energy levels should be carefully controlled to avoid overheating tissues.

3. EMS Safety Considerations

- EMS is also a safe and well-tolerated treatment when used appropriately. However, EMS should not be used on individuals with **heart conditions**, **epilepsy**, or **active cancer**, as the electrical impulses could interfere with the body's natural functions.
- It is important to ensure that the **electrode placement** is correct to prevent unnecessary muscle strain or injury. Training and supervision are key to effective and safe application.

Future of RF and EMS in Clinical Practice

The applications of RF and EMS therapies in clinical practice continue to expand. As research progresses, these technologies are becoming more refined, offering enhanced precision, higher levels of customization, and greater patient accessibility.

AI-driven protocols and **wearable devices** are expected to become more integrated into RF and EMS treatments, allowing for real-time monitoring and adjustment of therapeutic parameters based on patient response. This combination will likely lead to even more personalized treatment regimens, improving outcomes and minimizing risks.

Additionally, as healthcare systems move towards **cost-effective solutions** for muscle recovery and rehabilitation, **home-use devices** for RF and EMS will become more common, allowing patients to continue their therapies outside of clinical settings.

Conclusion

RF and EMS therapies are revolutionizing muscle recovery and rehabilitation in clinical practice. Their ability to promote tissue regeneration, reduce inflammation, enhance circulation, and stimulate muscle growth makes them invaluable tools in **physical therapy**, **sports medicine**, and **pain management**. Through real-world applications and case studies, we see the significant impact these therapies have on **patient recovery**, **muscle regeneration**, and **overall well-being**. With ongoing advancements in technology and the integration of **AI-driven optimization**, RF and EMS will continue to play a central role in the future of **bioelectric muscle therapies**, offering safe, effective, and non-invasive solutions for a variety of medical conditions.

Chapter 11: The Role of Sleep and Rest in Bioelectric Muscle Growth

One of the most fundamental yet often overlooked components of muscle growth and recovery is the role of **sleep and rest**. While **training** and **nutrition** are essential for stimulating muscle growth, the process of muscle recovery and the ultimate gains in strength, size, and function occur primarily during periods of rest, especially during sleep. In this chapter, we will explore the critical importance of sleep and rest in bioelectric muscle therapies, specifically how these periods support the bioelectric processes and hormonal balance that drive muscle regeneration. Additionally, we will discuss how **AI** can optimize rest cycles, integrating them with **RF** and **EMS** therapies to accelerate muscle recovery.

The Importance of Recovery

The body's muscle-building process relies on **recovery** after intense physical activity. During exercise, muscle fibers experience **microtears**, and the body's **anabolic processes** begin to repair and rebuild these fibers, making them stronger and larger. However, this process doesn't happen while working out; it primarily occurs during periods of **rest** and **sleep**.

The key to muscle growth is not the duration or intensity of the workout itself but the body's ability to repair the microdamage inflicted during training. Proper rest allows for several **physiological processes** that are critical for muscle recovery:

1. **Protein Synthesis**: Muscle growth occurs when protein synthesis exceeds muscle breakdown. Rest allows for the increased production of **amino acids** and **proteins**, essential for rebuilding muscle fibers.
2. **Cellular Repair**: The body's tissues are constantly in a state of repair. During rest, there is a surge in the production of **collagen**, which is essential for muscle regeneration and recovery from injury.
3. **Neurological Rebalance**: Rest and recovery periods allow the nervous system to **recover** from the stress of intense workouts. This is essential for proper **muscle activation** and **coordination** during subsequent training sessions.
4. **Reduction of Inflammation**: Sleep and rest help reduce inflammation caused by exercise-induced muscle damage. Proper recovery reduces the buildup of **cortisol**, a catabolic hormone that can lead to muscle breakdown if chronically elevated.

Hormonal Release During Sleep

While rest is critical for muscle recovery, **sleep** plays an even more pivotal role in the regulation of key **hormones** that contribute to muscle growth and regeneration. **Growth hormone (GH)**, **testosterone**, and **insulin-like growth factor (IGF-1)** all work together during sleep to enhance muscle synthesis, tissue regeneration, and fat metabolism.

1. **Growth Hormone (GH)**:

   GH is one of the most important hormones for muscle recovery. The majority of GH secretion occurs during **deep sleep stages**, specifically **slow-wave sleep (SWS)**. GH stimulates **protein synthesis**, helps **repair muscle tissue**, and promotes the regeneration of muscle fibers. This process is particularly important for athletes or individuals engaging in intensive physical activity.

2. **Testosterone**:

   Testosterone, a potent anabolic hormone, is also predominantly released during sleep. It enhances muscle hypertrophy by **increasing protein synthesis** and stimulating satellite cells that repair and grow muscle fibers. Testosterone also plays a key role in **reducing fat** and enhancing muscle strength.

3. **Insulin-Like Growth Factor (IGF-1)**:

   IGF-1 is released in response to GH and plays a crucial role in stimulating muscle regeneration and growth. It activates **satellite cells** that fuse with muscle fibers to promote muscle repair and hypertrophy. Adequate sleep supports the optimal release of GH and IGF-1, contributing to an effective recovery process.

4. **Cortisol Regulation**:

    Sleep also regulates the production of **cortisol**, a stress hormone that can have negative effects on muscle growth if elevated for prolonged periods. Cortisol inhibits **protein synthesis** and can lead to **muscle breakdown** if chronically high. Restful sleep lowers cortisol levels, allowing the body to remain in an **anabolic state** conducive to muscle recovery.

Integrating Rest with AI-Driven Muscle Growth

The integration of **artificial intelligence (AI)** with **bioelectric muscle therapies** like RF and EMS can take muscle growth and recovery to the next level by optimizing rest cycles. AI-driven systems can monitor data such as **sleep quality, muscle fatigue, recovery metrics**, and **hormonal levels**, enabling a more personalized and effective approach to rest and recovery.

1. **Sleep Monitoring and Optimization**:

    AI-powered wearables and smart devices can track sleep cycles, including **deep sleep stages** where most GH and testosterone release occurs. By analyzing data on sleep quality and quantity, AI can suggest specific **recovery protocols** or **treatment sessions** at the optimal times to maximize the regenerative effects of sleep. For instance, AI may recommend a gentle **EMS session** before bed to help relax muscles and promote deeper sleep, enhancing the overall recovery process.

2. **Recovery Planning**:

    AI can personalize recovery strategies based on data collected from workouts, treatments, and rest. If a user experiences **insufficient sleep** or signs of **muscle overtraining**, AI can adjust treatment protocols by increasing the intensity of RF therapy or EMS sessions during the day to facilitate more rapid muscle repair. This ensures that the muscles are receiving the right level of stimulation to enhance growth without overwhelming the recovery process.

3. **Optimal Rest Times**:

    AI-driven systems can also optimize **rest periods** by suggesting when to take breaks from intense training, based on real-time data about muscle fatigue, soreness, and recovery progress. These insights allow athletes and fitness enthusiasts to **avoid overtraining** while ensuring that rest is being utilized most effectively to promote muscle regeneration.

4. **AI-Enhanced Biofeedback**:

    AI systems that incorporate **biofeedback** from wearable devices and muscle sensors can help monitor muscle responses during rest. For example, AI can track muscle tension and identify signs of **overuse** or **under-recovery**, suggesting optimal times for rest or therapy. This continuous feedback loop helps fine-tune recovery protocols to maximize the balance between rest, training, and therapeutic interventions.

Optimizing Rest Cycles with RF and EMS Therapy

While rest is crucial, **RF therapy** and **EMS** can further enhance the restorative effects of sleep by promoting **circulation**, reducing **muscle stiffness**, and encouraging muscle regeneration even during rest periods. These therapies can be strategically applied to help the body remain in an **anabolic state** during recovery.

1. **RF Therapy for Enhanced Recovery**:

   RF therapy can stimulate **blood flow** and **collagen production**, helping muscles recover faster even during sleep. By applying RF therapy in the evening before sleep, individuals can improve **muscle tissue health** and reduce **inflammation** that might otherwise interfere with the restorative process during sleep. AI can help schedule these sessions based on sleep patterns and recovery needs.

2. **EMS for Pre-Rest Muscle Activation**:

   EMS can help activate muscles and prevent **muscle atrophy** during periods of rest, especially in individuals recovering from injuries or surgeries. EMS sessions before sleep can promote muscle relaxation and prepare muscles for the next day's activities, ensuring that recovery continues effectively even when the body is at rest.

Conclusion

The role of **sleep and rest** in muscle growth cannot be overstated. Rest is not just a passive process; it is an active, regenerative phase that allows the body to repair and grow muscle fibers. During sleep, the body optimally releases hormones like **growth hormone, testosterone**, and **IGF-1**, all essential for muscle regeneration. Integrating **AI-driven optimization** with **RF and EMS therapies** ensures that rest periods are used most effectively, accelerating recovery and enhancing the bioelectric signals involved in muscle growth.

As we continue to explore the relationship between **rest**, **sleep**, and bioelectric therapies, it is clear that **personalized recovery protocols** will be essential for maximizing muscle growth, strength, and overall performance. By harnessing the power of AI to monitor and optimize rest cycles, combined with the therapeutic benefits of RF and EMS, individuals can achieve faster, more efficient muscle regeneration while supporting their body's natural recovery processes. The future of muscle growth lies in balancing the physical stress of training with the restorative power of sleep and rest—enhanced by intelligent, data-driven systems that facilitate optimal recovery.

Chapter 12: The Future of AI and Bioelectric Muscle Growth

The integration of **Artificial Intelligence (AI)** with **bioelectric therapies** such as **Radiofrequency (RF) therapy** and **Electromagnetic Stimulation (EMS)** has already made significant strides in enhancing muscle growth, regeneration, and recovery. As these technologies continue to evolve, they promise to revolutionize the way we approach muscle health, fitness, and rehabilitation. This chapter will explore the emerging technologies in the field, the potential for integration with **wearable devices**, and the broader implications for athletes, fitness enthusiasts, and those recovering from injuries.

Emerging Technologies in AI, RF Therapy, and EMS

The future of **AI-driven bioelectric muscle growth** will rely on continuous advancements in **machine learning**, **data analytics**, and **bioelectric stimulation technologies**. AI systems will become increasingly sophisticated, enabling real-time optimization and personalized recovery protocols. As this technology matures, the ability to harness AI to drive muscle regeneration and growth will be unparalleled, potentially exceeding traditional training and rehabilitation methods.

1. **AI-Enhanced Data Collection and Analysis** As wearable devices and sensors continue to advance, they will become more adept at **tracking and analyzing** the body's response to exercise, therapy, and recovery. With AI's ability to process vast amounts of data, we will see muscle growth protocols tailored to individual needs, factoring in factors such as **age**, **genetics**, **training history**, and **sleep patterns**. AI will allow for continuous improvement of these protocols, adjusting the intensity, duration, and frequency of **RF therapy** and **EMS treatments** based on real-time feedback.

2. **Integration with Advanced Machine Learning Algorithms** AI will use **machine learning algorithms** to predict outcomes and recommend the best treatment paths for muscle growth. These algorithms will continuously learn from user data and improve the effectiveness of **bioelectric muscle therapy**. For example, the AI might recommend a change in the frequency of **RF energy pulses** or adjust the strength of **EMS** impulses based on muscle recovery metrics, pain levels, and overall progress.

3. **AI-Driven Personalization of Treatment Protocols** As AI systems become more powerful, they will offer even more **personalized treatment plans** for muscle growth and recovery. By analyzing individual muscle response and collecting data on workout performance, sleep quality, nutrition, and more, AI can create a precise bioelectric muscle treatment protocol. This would allow for optimized results that adapt to the user's progress over time, ensuring the most efficient and effective path to muscle growth and regeneration.

Integration with Wearable Devices

The **future of bioelectric muscle therapies** will also see deeper integration with **wearable devices** that monitor muscle performance and provide real-time biofeedback. These devices will allow individuals to track key metrics and adjust their routines accordingly, increasing the effectiveness of both **RF therapy** and **EMS**. Here are some possible ways these technologies will evolve:

1. **Wearable Muscle Stimulators and Sensors** Wearable devices will become increasingly common for both fitness enthusiasts and those undergoing rehabilitation. These devices will combine **muscle sensors** with **RF therapy** and **EMS technologies** to optimize muscle activation, reduce soreness, and prevent injury. For example, a **smart shirt** or **smart shorts** embedded with sensors could track muscle activation during workouts and adjust EMS impulses to target specific muscle groups that need recovery.
2. **Real-Time Biofeedback for Muscle Growth Optimization** AI will process the data from wearables to deliver **real-time biofeedback** to users, allowing them to adjust their workouts, rest, and therapy schedules. Devices will track the **intensity** of muscle contractions and **recovery progress**, delivering **real-time insights** on optimal recovery times, treatment frequencies, and areas of muscle tension that need more focus.

3. **Smart Recovery Systems** A wearable **smart recovery system** could be programmed to deliver targeted RF and EMS therapy automatically, based on collected data from the sensors. This system would analyze muscle soreness, fatigue levels, and recovery metrics, adjusting the therapy settings (frequency, intensity, and duration) to optimize muscle repair and growth. AI-driven **smart recovery wearables** could even synchronize with a user's sleep cycle, recommending recovery sessions before or after sleep to promote maximum **muscle regeneration**.

Bioelectric Muscle in Sports and Fitness

As AI, RF therapy, and EMS systems continue to evolve, athletes and fitness enthusiasts will see significant benefits from these technologies. The combination of AI-driven optimization, wearable devices, and bioelectric muscle therapies will provide more efficient ways to enhance **muscle strength**, **endurance**, and **recovery**.

1. **Enhanced Performance for Athletes** In the world of professional sports, **RF therapy** and **EMS** are already being used to **accelerate recovery** and improve **muscle endurance**. With AI technology, these therapies will become even more **tailored** to individual athletes, ensuring **optimal muscle activation** during training and **efficient recovery** after high-intensity workouts. Personalized treatment protocols will help athletes train harder, recover faster, and reduce the risk of injury.

2. **Sports Performance Monitoring** AI-powered wearables will monitor **athletic performance** in real time, tracking metrics such as muscle activation, recovery times, and muscle fatigue. This data will be used to recommend **RF therapy** and **EMS schedules** that complement training regimens, helping athletes build muscle more efficiently and recover without overtraining. The integration of **data analytics** and **machine learning** will make it easier to predict when athletes need recovery interventions and how to avoid muscle overuse.

3. **Rehabilitation for Injury Recovery** Athletes recovering from injuries will benefit greatly from the continued evolution of **bioelectric muscle therapies**. AI will tailor **rehabilitation plans** based on an athlete's injury history, muscle response, and performance data. Combining this with **RF therapy** and **EMS** will speed up the healing process, reduce inflammation, and promote tissue regeneration, allowing athletes to return to their sport faster and with fewer setbacks.

Beyond athletes, these **bioelectric muscle therapies** will have a broad societal impact, particularly in the field of **health and wellness**. By making advanced technologies accessible to the general public, the integration of **AI, RF therapy**, and **EMS** will provide an **affordable, effective approach** to muscle health.

1. **Enhanced General Wellness** With the increasing availability of **wearable bioelectric devices** and **home-use EMS and RF systems**, anyone looking to improve muscle tone, boost strength, or recover from physical activity can benefit. These systems, supported by **AI-based recommendations**, will provide individuals with the ability to enhance their muscle health from the comfort of their own homes.

2. **Improved Access to Rehabilitation** Bioelectric muscle therapies are a boon for those recovering from **injuries or surgeries**. The ability to use **RF therapy** and **EMS** at home, personalized by AI, will reduce the need for frequent clinic visits and make muscle rehabilitation more accessible. This could be a game-changer for individuals who live in areas with limited access to medical professionals or facilities.

3. **Chronic Pain and Musculoskeletal Disorder Management** For individuals suffering from chronic pain or musculoskeletal disorders like arthritis or fibromyalgia, the integration of **bioelectric muscle therapies** will provide a non-invasive, drug-free approach to treatment. Personalized AI-driven recovery protocols will help manage pain, improve mobility, and reduce dependence on medications.

Conclusion

The future of **AI-driven bioelectric muscle growth** represents a dramatic shift in how we approach muscle recovery, performance, and rehabilitation. As **AI**, **RF therapy**, and **EMS technologies** continue to evolve and integrate with wearables and other smart systems, the possibilities for enhancing muscle health will expand exponentially. Whether you are an elite athlete, a fitness enthusiast, or someone recovering from injury, these advanced technologies will offer new, personalized, and highly effective ways to optimize muscle growth, improve performance, and accelerate recovery. The future of **bioelectric muscle therapies** is bright, and with further advancements in **AI** and **bioelectric stimulation**, we can expect to see even greater breakthroughs in muscle regeneration, growth, and health in the years to come.

Chapter 13: Ethical Considerations in Bioelectric Muscle Therapy

As with any groundbreaking technology, the use of **bioelectric muscle therapy**—combining **artificial intelligence (AI)**, **radiofrequency (RF)** therapy, and **electromagnetic stimulation (EMS)**—brings forth a series of ethical concerns that must be addressed to ensure responsible application and regulation. While these technologies promise immense benefits for muscle regeneration, growth, and recovery, they also raise critical questions about privacy, access, safety, and fairness. In this chapter, we will explore the key ethical issues surrounding the use of AI, RF, and EMS in muscle therapy, particularly in medical, fitness, and non-medical settings.

Ethical Issues with AI in Medicine

AI-driven technologies are revolutionizing healthcare by providing **personalized treatment plans**, enhancing the precision of medical devices, and enabling real-time monitoring and adjustment of therapy protocols. However, the use of AI in medicine—particularly in muscle regeneration and growth—requires a careful evaluation of the following ethical concerns:

Privacy and Data Security

personal data

personalizing treatments

privacy

data encryption

secure storage protocols

Informed Consent

2. **Bias and Fairness**: AI systems are only as good as the data they are trained on. If an AI system is trained on a biased dataset, it could lead to **discriminatory outcomes**. For example, muscle growth protocols designed by AI may work better for individuals from certain demographic groups (e.g., age, gender, or ethnicity) and may be less effective for others. Ensuring **diverse and representative datasets** is crucial for achieving fair and unbiased treatment recommendations.
3. **Autonomy vs. AI Control**: AI-driven therapy systems that monitor and adjust treatments in real-time raise the question of **autonomy**. Should AI have the power to alter treatment protocols without human oversight? While AI can optimize muscle therapy and recovery, it is important to ensure that human healthcare professionals remain in control of critical decisions. Ethical guidelines must be established to ensure **human oversight** while still allowing AI to enhance therapy outcomes.

Bioelectric Stimulation in Non-Medical Settings

While bioelectric muscle therapies, such as RF and EMS, have shown great promise in medical settings, their use outside clinical environments—such as in gyms, fitness centers, or at-home treatment systems—raises additional ethical concerns.

Performance Enhancement

RF therapy

EMS

performance enhancement

cosmetic

performance-based enhancement

fair competition

Sports Integrity

arms race

regulated

Consumer Protection

at-home bioelectric muscle devices

consumer protection

RF therapy

EMS

clear guidelines

safety protocols

Lack of Supervision

third-party certifications

Accessibility and Equity

impressive benefits

AI-driven therapy systems

wearable devices

health inequalities

Addressing Disparities

public health initiatives

subsidies

universal healthcare integration

Balancing Innovation and Ethics

The potential of **bioelectric muscle therapies**—when combined with AI, RF, and EMS—offers immense benefits for muscle growth, regeneration, and recovery. However, these innovations must be balanced with ethical considerations to prevent harm, ensure fairness, and protect individual rights. Here are a few key strategies to achieve this balance:

1. **Developing Ethical Guidelines and Regulations**: Establishing comprehensive **ethical guidelines** and **regulations** will be essential for the responsible implementation of bioelectric muscle therapies. These guidelines should cover everything from **data privacy** and **informed consent** to ensuring **fair access** and **safety standards** for consumers and patients.
2. **Collaborative Efforts**: It is crucial to foster **collaboration between stakeholders**—including **scientists, regulatory bodies, clinicians**, and **technology developers**—to ensure that AI-driven bioelectric muscle systems are developed responsibly and are accessible to all. Such collaborations can also help in creating frameworks to evaluate the **long-term effects** and **safety** of these therapies.
3. **Public Education and Awareness**: Educating the public on the ethical implications and responsible use of bioelectric therapies will be essential to ensure that people use these systems in a way that benefits their health and wellbeing. Public awareness campaigns could address common concerns such as **misuse**, **overuse**, and **safety protocols** for at-home systems.

4. **Continuous Monitoring and Feedback**: Ethical oversight will require **continuous monitoring** of bioelectric muscle systems and AI algorithms, ensuring that these technologies evolve in ways that prioritize the **safety**, **privacy**, and **fairness** of users. Regulatory bodies should track the long-term impacts of these therapies on muscle health and adjust guidelines as necessary to ensure that ethical standards are met.

Conclusion

Bioelectric muscle therapy represents a transformative leap in how we understand muscle growth, regeneration, and recovery. By combining **AI**, **RF therapy**, and **EMS**, these technologies hold the potential to revolutionize both clinical and non-clinical treatments. However, as with any powerful technology, it is vital to ensure that **ethical considerations** are not overlooked. Issues such as **privacy**, **data security**, **accessibility**, **equity**, and **sports fairness** must be addressed thoughtfully and proactively to ensure that the benefits of these therapies are realized without causing harm.

The responsible use of bioelectric muscle therapies requires careful oversight and the development of ethical frameworks that ensure these innovations improve the lives of individuals in a fair and just manner. By balancing technological progress with ethical responsibility, we can unlock the full potential of AI-driven **bioelectric muscle systems**, transforming muscle recovery and performance in ways that benefit society as a whole.

Chapter 14: Building the Bioelectric Muscle System

The future of muscle growth and regeneration is poised for revolution with the advent of technologies like **artificial intelligence (AI)**, **radiofrequency (RF) therapy**, and **electromagnetic stimulation (EMS)**. These systems offer groundbreaking potential for optimizing muscle recovery, performance, and rehabilitation. However, to truly harness the power of bioelectric muscle technologies, we must understand how to **build**, **integrate**, and **personalize** these systems for maximum effectiveness.

In this chapter, we will explore the components of a **bioelectric muscle system**, how to design a **customized solution** tailored to individual needs, and predict the **future trends** in bioelectric muscle therapies. This chapter will serve as a blueprint for professionals, entrepreneurs, and innovators interested in developing bioelectric muscle systems that can be used across various settings—from rehabilitation clinics to home-use devices.

Components of a Bioelectric Muscle System

A successful bioelectric muscle system involves a sophisticated blend of hardware, software, and clinical expertise. Let's break down the essential components of such a system:

Hardware

- **RF Therapy Devices**: These devices emit radiofrequency energy, targeting tissues deep within the body. The system must include precise RF emitters and modulators that can adjust wave frequencies, intensity, and duration based on the patient's specific needs.
- **EMS Units**: EMS systems utilize electrical impulses to stimulate muscle contraction. The units should be designed to deliver controlled electrical signals to muscle fibers, with adjustable parameters like frequency, intensity, and duration of stimulation.
- **Sensors**: Integrated sensors, such as **electromyography (EMG)** sensors, measure muscle activity, fatigue, and progress. These sensors help ensure that treatments are optimized and targeted at the right muscle groups.
- **Wearable Devices**: Devices like smart bands, straps, or patches can be used to deliver RF or EMS therapy on a more portable basis, providing ongoing therapy and monitoring.

Software

- **AI Algorithms**: AI plays a central role in bioelectric muscle systems, offering the ability to analyze real-time data from sensors, provide recommendations, and adjust treatment parameters. AI-driven software can optimize RF and EMS treatment protocols based on individual muscle response, injury history, and specific fitness goals.
- **User Interface (UI)**: The software should have an intuitive user interface that allows users (whether healthcare professionals or patients) to monitor progress, adjust settings, and track outcomes. This may include dashboards that visualize muscle activity, growth, and recovery trends over time.
- **Data Analytics Tools**: The system should have built-in data analytics tools that enable users to assess muscle regeneration and performance improvements. These tools can aggregate data from multiple sessions, offering insight into how the therapy is impacting muscle health.

Clinical Expertise

- **Medical Oversight**: Any bioelectric muscle system, particularly one intended for clinical use, requires oversight by medical professionals. These experts ensure that the therapy is administered safely and effectively, monitoring patient responses, and adjusting protocols as needed.
- **Personalized Treatment Plans**: Bioelectric muscle systems should provide tailored therapy, taking into account factors like muscle health, injuries, age, and activity levels. **Clinical guidance** ensures the patient is receiving the most effective protocol.

Designing a Personalized System

One of the core strengths of bioelectric muscle systems lies in their ability to provide personalized therapy. To design a customized system, the following steps should be considered:

Initial Assessment

- **Physiological Evaluation**: Before using a bioelectric muscle system, a comprehensive **physiological assessment** must be conducted. This includes measuring muscle strength, fiber density, range of motion, and identifying any pre-existing injuries. The system can then be calibrated to meet the individual's muscle health and recovery needs.
- **Health and Fitness Goals**: Each individual's goals must be clearly defined—whether they seek to build muscle mass, recover from injury, or enhance athletic performance. The system should be able to adjust its parameters based on these goals.

Customization of Treatment Protocols

- **Adjusting Frequency and Intensity**: The system should allow for **dynamic adjustments** to frequency, intensity, and duration based on feedback from sensors. For example, an athlete might require higher-intensity EMS to stimulate deeper muscle contractions, while an individual recovering from an injury may benefit from lower-frequency settings.
- **Growth Hormone Stimulation**: For users looking to enhance muscle recovery and growth, the system should integrate protocols that stimulate the release of **growth hormone** through RF therapy, further accelerating muscle regeneration.
- **Ongoing Monitoring**: Personalized systems should continuously monitor the effectiveness of treatments through integrated sensors. Data collected during each session can be fed back into the system to fine-tune future treatments, ensuring that the protocol evolves based on real-time responses.

Patient Feedback

- **Biofeedback Mechanisms**: Incorporating **biofeedback mechanisms** allows the system to respond to the individual's body signals during treatment. For example, if the system detects muscle fatigue, it can automatically adjust the stimulation intensity, ensuring that the muscle does not undergo excessive stress or injury.
- **User-Driven Input**: Users should also be able to manually adjust settings, providing feedback on comfort levels or perceived effectiveness. This ensures that users remain actively involved in their treatment process.

Future Trends in Bioelectric Muscle Systems

As the field of bioelectric muscle therapy continues to evolve, several key trends are emerging that promise to further enhance the efficacy and accessibility of these systems:

Integration with Wearables

- The future of bioelectric muscle therapy will likely involve the integration of wearables like **smart clothing** or **skin patches**. These devices will provide real-time, continuous monitoring of muscle activity, fatigue, and recovery, while also delivering treatment without the need for bulky equipment.
- **Wearables** will also enable remote monitoring by healthcare providers, allowing for adjustments in therapy protocols without requiring patients to visit a clinic.

Cloud-Based Systems

- **Cloud technology** is expected to play a major role in bioelectric muscle systems. By leveraging cloud computing, patient data can be securely stored and analyzed, and therapists can access treatment plans, data trends, and patient progress from anywhere.
- Cloud-based AI systems could provide **automated treatment adjustments**, optimizing therapy in real-time and sharing insights across multiple platforms to help healthcare providers make better-informed decisions.

AI-Enhanced Recovery Algorithms

- AI algorithms will continue to improve, analyzing not only muscle responses but also **long-term trends** in recovery. For example, AI could predict future injuries or muscle imbalances before they occur, enabling proactive interventions.
- Future systems may also be able to integrate with **genomic data**, allowing for **personalized muscle growth protocols** that are tailored not just to current fitness levels but also genetic predispositions.

Global Accessibility and Affordability

- As bioelectric muscle therapies gain traction, there will be a concerted push to make these technologies more **affordable** and **accessible** to the general public. The development of more cost-effective devices, supported by scalable cloud-based AI systems, will help democratize access to these treatments.
- Partnerships with **health insurance companies** and **public health initiatives** could enable bioelectric muscle therapies to be more widely adopted, even in lower-income regions.

Conclusion

Building an effective bioelectric muscle system requires a seamless integration of hardware, software, and clinical expertise. The key to success lies in designing personalized treatment protocols that are both adaptable and precise. As technological advancements continue to shape this field, future bioelectric muscle systems will provide even greater levels of **customization**, **optimization**, and **efficiency**. These systems will help individuals not only recover faster but also achieve enhanced muscle performance and regeneration, revolutionizing how we approach muscle health, fitness, and recovery.

The potential of bioelectric muscle therapy is vast. With careful design and continual improvement, these systems will not only enhance muscle regeneration but also push the boundaries of human physical performance, ultimately leading to healthier, stronger individuals across the globe.

Chapter 15: Overcoming Challenges in Bioelectric Muscle Therapy

The growing potential of bioelectric muscle therapy—through AI-driven optimization, radiofrequency (RF) therapy, and electromagnetic stimulation (EMS)—presents an exciting frontier for muscle growth, recovery, and regeneration. However, like any emerging field, it faces a variety of challenges that must be overcome to achieve widespread adoption and maximize its benefits. In this chapter, we will explore these challenges, including resistance to new technologies, technological barriers, and how to optimize patient outcomes. Understanding and addressing these obstacles is crucial for the continued development and success of bioelectric muscle therapy.

Resistance to New Technologies

Introducing new technologies into any field, particularly in healthcare, is often met with skepticism and resistance. This challenge is no different when it comes to bioelectric muscle therapy. The reluctance to adopt AI, RF, and EMS therapies can stem from several factors:

1. **Lack of Awareness and Education**: Many healthcare professionals and patients are unfamiliar with the capabilities and advantages of bioelectric muscle therapies. This lack of understanding can lead to hesitation in incorporating these technologies into regular treatment plans.

2. **Traditional Methods vs. New Approaches**: Healthcare practitioners who are accustomed to traditional therapeutic techniques may be resistant to adopting new methods. RF and EMS treatments, for instance, can seem unconventional compared to well-established modalities like physical therapy or exercise-based rehabilitation.

3. **Trust in Artificial Intelligence**: The use of AI to personalize and optimize treatments is a major breakthrough in muscle therapy, but some individuals still find it difficult to trust AI-driven systems. Concerns about the accuracy, reliability, and safety of AI-based decisions may prevent professionals from embracing these systems.

4. **Cultural and Institutional Resistance**: In some medical and rehabilitation environments, there may be institutional inertia or cultural resistance to adopting cutting-edge technologies. This resistance can be particularly strong in conservative medical institutions or where regulatory bodies have not yet fully endorsed these new therapies.

Overcoming Resistance:

- **Education and Training**: For healthcare professionals, a robust training program that explains the science, benefits, and practical applications of AI, RF, and EMS therapies is essential. This will help reduce resistance and increase confidence in these technologies.
- **Success Stories and Case Studies**: Demonstrating the efficacy of bioelectric therapies through well-documented case studies and patient success stories is a powerful way to shift perceptions. Real-world examples of improved patient outcomes will help clinicians and patients alike understand the value of these therapies.
- **Collaboration with Innovators**: Collaborating with leading researchers and developers in the field of bioelectric muscle therapy can help build trust and credibility. Hospitals, rehabilitation centers, and fitness clinics can benefit from forming partnerships with technology providers to pilot bioelectric systems.

Technological Barriers

While bioelectric muscle therapies offer significant potential, there are several technological challenges that need to be addressed to improve their accessibility, effectiveness, and ease of use:

1. **Sensor Technology**: Accurate and reliable sensors are critical for monitoring muscle activity and providing real-time feedback. Current sensors may not always offer the precision required to monitor subtle changes in muscle performance or recovery. Inconsistent sensor readings can lead to ineffective treatments.
2. **Data Integration and Analysis**: Collecting data from multiple sources—sensors, RF, EMS units, and AI systems—requires robust data integration platforms. These platforms must process large volumes of real-time data to ensure that bioelectric systems can adjust protocols based on muscle responses, fatigue levels, and progress.
3. **Device Portability and Usability**: While some bioelectric muscle systems are designed for clinical use, there is a growing need for portable, user-friendly devices that can be used at home or on the go. Ensuring that these devices are intuitive and easy to use for non-experts is crucial for adoption in personal health settings.
4. **Safety and Comfort**: As with any new technology, ensuring the safety and comfort of patients is paramount. RF and EMS treatments must be calibrated to avoid over-stimulation or potential injury. Additionally, devices must be designed for long-term wearability without causing discomfort or irritation.

Addressing Technological Barriers:

- **Advancements in Sensors**: The development of more sensitive, reliable sensors is essential for improving the effectiveness of bioelectric muscle systems. Wearable sensors, such as those integrated into clothing or adhesive patches, must be capable of providing accurate readings of muscle activity without compromising comfort.
- **AI-Driven Data Processing**: AI can be used to streamline data processing by filtering out noise and providing actionable insights. As AI continues to evolve, it will be capable of offering more refined analyses and adapting treatments in real-time to optimize muscle regeneration.
- **User-Centric Design**: Focusing on the user experience is essential for ensuring that bioelectric systems are accessible to a broader population. Simplifying device interfaces and enhancing comfort through ergonomic design will make these technologies more appealing and usable for patients and consumers alike.
- **Safety Protocols and Monitoring**: Standardized safety protocols for RF and EMS treatments need to be developed and implemented. This includes clear guidelines for treatment intensity, duration, and frequency to minimize the risk of adverse effects.

Optimizing Patient Outcomes

The ultimate goal of bioelectric muscle therapy is to improve patient outcomes by accelerating muscle recovery, enhancing growth, and reducing the risk of injury. However, several factors influence the success of these treatments:

1. **Patient Variability**: Each patient is unique, and factors such as age, genetics, medical history, and activity level will all impact how they respond to bioelectric therapy. For example, athletes may require more intensive stimulation than individuals recovering from surgery or injury.
2. **Personalization of Treatment**: One of the strengths of bioelectric muscle therapy is its ability to provide personalized treatment. However, creating truly customized protocols requires accurate data and sophisticated AI algorithms that can continuously adapt to a patient's evolving needs.
3. **Consistency of Use**: Achieving optimal results from bioelectric muscle therapy requires consistent and regular use. Patients must be committed to adhering to prescribed treatment plans, which can be challenging if they don't experience immediate results or if treatments are not easily accessible.
4. **Integration with Other Therapies**: Bioelectric muscle therapy should not be viewed as a standalone treatment but as part of a holistic recovery or growth plan. It works best when integrated with other therapeutic approaches, such as physical therapy, strength training, and proper nutrition.

Optimizing Patient Outcomes:

- **Continuous Monitoring and Adjustment**: AI-driven systems can help optimize patient outcomes by adjusting treatment parameters in real-time based on muscle feedback. By continuously monitoring muscle activity and progress, the system can fine-tune its settings to enhance recovery and performance.
- **Patient Engagement and Education**: For patients to fully benefit from bioelectric muscle therapy, they must be engaged in the process. Providing clear educational materials about the science behind the therapy and its benefits will encourage patients to stay committed to their treatment plans.
- **Multi-Therapeutic Approach**: Combining bioelectric muscle therapy with complementary treatments such as nutrition, sleep optimization, and strength training will help maximize results. A multi-faceted approach addresses all aspects of muscle recovery and growth, leading to better long-term outcomes.

Conclusion

Overcoming the challenges in bioelectric muscle therapy is crucial for ensuring the widespread adoption and success of these technologies. By addressing resistance to new technologies, overcoming technological barriers, and optimizing patient outcomes, we can unlock the full potential of bioelectric systems to transform muscle recovery, growth, and rehabilitation. With continued innovation and a focus on patient-centered care, bioelectric muscle therapy will become a cornerstone of modern healthcare, fitness, and athletic performance. The future of muscle regeneration is bright, and with these solutions in place, the world will witness a new era of muscle health and recovery.

Chapter 16: RF and EMS for Injury Prevention

In the realm of muscle growth and recovery, injury prevention is an area of paramount importance. Muscular injuries, whether from overuse, strain, or trauma, can hinder progress, slow recovery times, and in some cases, result in long-term damage. With the advancements in bioelectric muscle therapies, specifically radiofrequency (RF) therapy and electromagnetic stimulation (EMS), there is an emerging opportunity to not only repair muscle damage but to prevent injuries before they occur. In this chapter, we will explore how RF and EMS technologies can be applied to prevent muscle strain and overuse injuries, how electrical stimulation supports tissue repair, and how injury prevention programs are being successfully integrated into both clinical settings and personal training regimens.

Preventing Muscle Strain and Overuse Injuries

Muscle strain is one of the most common injuries affecting athletes, fitness enthusiasts, and individuals performing repetitive movements in daily life. Overuse injuries, in particular, result from continuous strain on muscles without adequate recovery time, leading to microtears and inflammation. The need for proactive injury prevention has never been more significant, and bioelectric therapies like RF and EMS are at the forefront of these efforts.

RF Therapy for Muscle Conditioning

- **Increased Circulation**: RF energy stimulates vasodilation, which improves blood flow and oxygen delivery to muscle tissues, allowing for quicker nutrient delivery and toxin removal. Improved circulation ensures that muscles remain well-nourished and can withstand greater stress during physical activity, reducing the risk of muscle strain.
- **Enhanced Collagen Production**: Collagen is a critical component of muscle and connective tissues. RF therapy stimulates fibroblast activity, which is responsible for collagen synthesis. Stronger, more elastic muscles and tendons are less prone to tearing and overuse injuries.

EMS for Muscle Activation

- **Pre-Activation of Muscles**: EMS can be used to activate muscles before an intense workout or physical activity, ensuring they are properly warmed up and prepared for the stresses ahead. This pre-activation prepares the muscle fibers, reducing the chance of sudden overstretching or injury.
- **Improved Muscle Endurance**: EMS aids in improving muscle stamina and endurance by working deeper muscle fibers that might not be fully activated during regular exercise. The increased endurance helps prevent fatigue-induced injuries during long-duration activities.

Strengthening Supporting Muscles

Role of Electrical Stimulation in Tissue Repair

Electrical stimulation therapies—both RF and EMS—are not only effective in injury prevention but also serve as powerful tools in accelerating tissue repair. Whether it's for rehabilitating a prior injury or ensuring that tissue remains strong and healthy, the application of electrical stimulation can significantly improve recovery times and quality of healing.

RF Therapy for Accelerated Healing

- **Reduction in Inflammation**: Inflammation is a natural part of the healing process, but chronic or excessive inflammation can slow recovery and lead to further damage. RF therapy has been shown to reduce inflammatory markers, allowing the tissue to heal more efficiently and preventing chronic conditions from setting in.
- **Enhanced Cellular Regeneration**: By stimulating the production of collagen and elastin, RF therapy promotes the formation of new tissue and accelerates the recovery of muscle fibers. The faster the tissue regenerates, the sooner the body can return to optimal performance levels.

EMS for Tissue Repair

- **Increased Oxygenation and Nutrient Delivery**: The contractions triggered by EMS improve the oxygen and nutrient supply to injured areas, allowing tissues to recover more rapidly.
- **Muscle Relaxation**: EMS can also be used to relax overly tight muscles, reducing spasms and discomfort associated with injuries. This relaxation effect improves circulation and helps prevent further strain on damaged areas.

Case Examples: Successful Injury Prevention Programs Using Bioelectric Technologies

Several real-world applications demonstrate how RF and EMS are being used successfully for injury prevention, particularly in professional sports, rehabilitation centers, and personal fitness regimens.

1. **Professional Sports Teams**: In professional sports, athletes often face the threat of overuse injuries due to the intense physical demands of their training and competition schedules. Many teams now incorporate RF and EMS therapies as part of their injury prevention programs. For example, EMS is used before training sessions to activate muscles and enhance performance, while RF therapy is used post-training to aid in muscle recovery and prevent strains.

2. **Rehabilitation Centers**: In clinical settings, physical therapists utilize RF and EMS technologies for patients recovering from musculoskeletal injuries. By incorporating these therapies early in the rehabilitation process, therapists can improve muscle function, reduce scar tissue formation, and strengthen injured areas, ultimately helping patients return to their daily activities more quickly and with reduced risk of re-injury.

3. **Home Use Devices**: Increasingly, consumers are gaining access to portable RF and EMS devices designed for home use. These devices allow individuals to incorporate muscle conditioning and injury prevention into their daily routines. Whether it's pre-workout muscle activation using EMS or post-workout recovery with RF therapy, these devices empower users to take control of their muscle health.

Conclusion

Injury prevention is a key component of any muscle growth and recovery program. The integration of RF therapy and EMS into both rehabilitation and prevention strategies offers tremendous benefits by reducing the risk of strain, improving muscle endurance, and accelerating tissue repair. These technologies not only help in maintaining peak performance levels but also ensure that muscles remain resilient and capable of handling the stresses placed on them.

As the understanding and adoption of bioelectric muscle therapies continue to expand, the role of RF and EMS in injury prevention will become even more central to the health and fitness industry. Whether used by professional athletes, rehabilitation patients, or fitness enthusiasts, these advanced technologies are helping to push the boundaries of muscle performance and recovery, all while minimizing the risk of injury. With the right application, RF and EMS therapies will continue to evolve and play a vital role in the future of muscle health.

Chapter 17: Bioelectric Muscle in Aging and Regenerative Medicine

As we age, our muscles undergo significant changes that can impact both their size and function. These age-related declines in muscle mass and strength can lead to conditions such as sarcopenia, a progressive loss of muscle tissue that can severely affect mobility and overall quality of life. However, emerging technologies like artificial intelligence (AI), radiofrequency (RF) therapy, and electromagnetic stimulation (EMS) are showing promise in combating these effects and providing solutions for muscle regeneration in aging individuals. In this chapter, we will explore how bioelectric muscle therapies can address age-related muscle loss, the potential for reversing sarcopenia, and the exciting possibilities of combining these therapies with stem cell treatments for regenerative medicine.

How Aging Affects Muscle Regeneration

Aging is associated with a variety of physiological changes that impact muscle health, including a decrease in muscle fiber size, a reduction in muscle stem cells, and a decline in satellite cell activity. Satellite cells play a vital role in muscle repair and regeneration by aiding in the formation of new muscle fibers. As these cells become less active with age, the body's ability to repair muscle damage decreases, making muscles more prone to injury and less capable of regeneration. Additionally, older adults often experience a decrease in the production of growth hormone (GH) and testosterone, both of which are crucial for muscle repair and protein synthesis.

Some of the primary effects of aging on muscle tissue include:

1. **Loss of Muscle Mass (Sarcopenia)**: Sarcopenia, the progressive loss of muscle mass, is one of the most common conditions in older adults. It contributes to frailty, weakness, and reduced functionality, making it difficult for aging individuals to perform everyday tasks.
2. **Decreased Muscle Strength**: Along with muscle mass, muscle strength declines with age. This makes it harder for older adults to maintain an active lifestyle and can increase the risk of falls and other injuries.
3. **Slow Recovery from Exercise**: Aging muscles take longer to recover from exercise due to the decreased ability of muscle cells to repair themselves effectively. The slowed recovery process also increases the likelihood of overuse injuries in older adults.
4. **Reduced Regenerative Capacity**: As we age, the capacity of muscles to regenerate after injury is reduced. This leads to slower healing times and a greater risk of permanent damage following trauma or strain.

Using Bioelectric Muscle Technologies to Reverse Age-Related Muscle Loss

AI, RF therapy, and EMS are poised to significantly alter the way we address muscle degeneration and aging. These technologies offer several potential benefits for older adults seeking to maintain or even regain muscle mass and function.

AI-Driven Muscle Regeneration

- **Personalized Muscle Training Programs**: AI can integrate data from wearable devices to monitor the user's muscle strength and endurance, then adjust workout or recovery protocols accordingly.
- **Biofeedback Systems**: Real-time biofeedback, which continuously adjusts the treatment based on muscle response, ensures maximum muscle activation without overstraining, helping older individuals safely improve muscle function.

RF Therapy for Rejuvenating Muscle Tissue

- **Collagen Synthesis**: RF therapy stimulates collagen production, an essential protein that helps to restore muscle tissue and improves elasticity, thus reducing the risk of injury in aging muscles.
- **Improved Circulation**: The deep heating effect of RF therapy promotes increased blood flow, which helps to nourish muscle tissues with oxygen and nutrients, accelerating the regeneration process.

EMS for Reversing Muscle Atrophy

- **Muscle Strengthening**: EMS helps to strengthen the muscles by engaging fibers that are often neglected during regular exercise. For older adults, this can be crucial in regaining lost strength and mobility.
- **Prevention of Further Muscle Loss**: Regular EMS treatments can help prevent the onset of further muscle deterioration by maintaining muscle activity and preventing the weakening of fibers.

Potential in Stem Cell Therapy

An exciting area of research lies in the combination of bioelectric muscle therapies with regenerative techniques, such as stem cell therapy. Stem cells have the unique ability to differentiate into various types of cells, including muscle cells. When combined with RF therapy or EMS, stem cells can be guided to repair and regenerate damaged muscle tissues, offering the potential for reversing some of the debilitating effects of aging on muscles.

1. **Stem Cell Activation with RF and EMS**: RF and EMS treatments can stimulate the activation and differentiation of stem cells located in muscle tissue. By providing a bioelectric signal, these therapies can encourage stem cells to become muscle cells, increasing muscle regeneration and improving tissue repair after injury or aging-related degeneration.
2. **Stem Cells for Sarcopenia Treatment**: Researchers are exploring the use of stem cell injections to restore muscle mass in aging adults. When combined with RF and EMS, this approach may offer even greater therapeutic benefits, providing a non-invasive method for rejuvenating aged muscle tissue.

Case Studies: Real-World Applications in Aging and Regenerative Medicine

Several case studies highlight the effectiveness of bioelectric muscle therapies in addressing the challenges of aging and muscle degeneration.

1. **Clinical Trials in Older Adults**: Clinical trials have demonstrated that RF and EMS therapies can significantly improve muscle strength and recovery times in older adults. One such trial involved patients suffering from sarcopenia who received a combination of EMS and RF therapy. Results showed increased muscle mass, reduced muscle stiffness, and improved overall strength after several weeks of treatment.
2. **Success Stories in Physical Rehabilitation**: Older individuals recovering from surgeries or fractures have found significant benefits from incorporating RF therapy and EMS into their rehabilitation programs. Patients who received these therapies were able to regain muscle mass and strength faster than those who only participated in traditional physical therapy.

Conclusion

The effects of aging on muscle health can be debilitating, but with the advent of bioelectric muscle therapies, there is newfound hope for older adults facing muscle loss and weakness. AI-driven treatments, RF therapy, and EMS offer promising solutions to reverse the decline in muscle mass and strength associated with aging. Additionally, combining these therapies with regenerative medicine, including stem cell treatments, opens up exciting possibilities for rejuvenating muscle tissue and improving mobility in older adults.

As research and technology continue to evolve, bioelectric muscle therapies will play an increasingly important role in regenerative medicine, offering more effective and accessible treatments for age-related muscle loss. By embracing these innovative therapies, older adults can not only slow the progression of sarcopenia but also regain the strength, flexibility, and independence that are critical to maintaining a healthy and active lifestyle.

Chapter 18: Cost-Effective Solutions for Bioelectric Muscle Growth

As the field of bioelectric muscle therapy continues to evolve, one of the most significant challenges is making these advanced technologies accessible and affordable for a broad audience. While AI-driven optimization, radiofrequency (RF) therapy, and electromagnetic stimulation (EMS) offer powerful benefits for muscle growth and regeneration, the high costs associated with specialized equipment and treatments often limit their availability. In this chapter, we explore ways to make bioelectric muscle therapies more cost-effective, leveraging innovative solutions like AI and cloud-based systems, as well as promoting accessible options for commercial and home use.

Affordable Technology: Making RF and EMS More Accessible

One of the primary hurdles to widespread adoption of RF and EMS therapies is the initial investment required for the necessary technology. High-end medical-grade equipment used in clinical settings can be prohibitively expensive, limiting these treatments to high-end facilities and well-funded research labs. However, there are several strategies being explored to reduce costs and make these therapies more widely available.

Miniaturization of Devices

- **Portable EMS Devices**: EMS devices, often marketed for personal use, can be employed in-home therapy, offering a low-cost alternative for muscle strengthening and rehabilitation. With advancements in wireless technology, users can adjust the intensity and frequency of stimulation through mobile applications, making it a personalized and user-friendly solution.
- **Compact RF Units**: Many smaller RF units are now available for home use. These devices operate on similar principles as those used in clinical settings but at a fraction of the cost. While the power output and depth of penetration may differ from professional equipment, they are still effective in promoting muscle healing, enhancing circulation, and reducing muscle tension.

Affordable Wearable Devices

- **Smart EMS Wearables**: These wearables use lightweight, flexible electrodes that send electrical impulses to muscles, helping to promote growth and recovery. They can be worn throughout the day and adjusted to target specific muscle groups, offering a customizable solution for ongoing therapy.
- **RF Therapy Wearables**: RF therapy can also be incorporated into wearable products like patches, bands, or pads, which emit controlled RF energy to stimulate the muscle and improve recovery. These products are user-friendly and convenient, making it easier for individuals to benefit from bioelectric therapies without needing to visit a clinic.

AI and Cloud-Based Systems: Reducing Costs Through Automation and Integration

AI and cloud-based solutions hold great promise for reducing the cost of bioelectric muscle therapies by automating and optimizing the treatment process. Through AI-driven systems, personalized treatments can be delivered at a fraction of the cost of traditional in-person consultations. By leveraging cloud computing, data processing, and real-time monitoring, bioelectric muscle therapy can become more affordable, personalized, and accessible.

AI-Optimized Treatment Plans

Data-Driven Insights

Cloud-Based Systems for Remote Access

Affordable AI-Powered Platforms

Commercial and Home Use Solutions: Expanding Accessibility

While professional-grade bioelectric muscle systems can be expensive, there are now numerous options available for consumers seeking affordable at-home solutions. These products provide similar benefits to clinical devices, but with a focus on ease of use and cost-effectiveness.

Over-the-Counter EMS and RF Devices

EMS Belts and Patches

Subscription Models for At-Home Therapy

Conclusion

The high costs associated with bioelectric muscle therapies should not be a barrier to their widespread use. Through technological advancements such as miniaturization, wearables, AI optimization, cloud-based systems, and affordable over-the-counter solutions, bioelectric muscle therapies are becoming more accessible and cost-effective for a broader audience. By reducing the cost of RF and EMS systems, these therapies can reach individuals in both clinical and home settings, providing valuable tools for muscle growth, recovery, and injury prevention.

As these technologies continue to advance, we can expect further innovations that will make bioelectric muscle therapies even more affordable and widely available. This democratization of technology will empower more people to take control of their muscle health, enhancing both quality of life and physical performance across all age groups.

Chapter 19: Real-World Applications and Success Stories

The fusion of artificial intelligence (AI), radiofrequency (RF) therapy, and electromagnetic stimulation (EMS) has the potential to revolutionize muscle regeneration, growth, and recovery. These technologies, though still evolving, are already making significant strides in both clinical and non-clinical settings. From professional athletes leveraging bioelectric muscle systems for enhanced performance to individuals experiencing remarkable muscle recovery transformations, the impact of these therapies is becoming increasingly evident. This chapter will delve into real-world applications and showcase success stories where bioelectric muscle therapies have made a difference in muscle growth, recovery, and overall well-being.

Professional Athletes and Bioelectric Muscle Growth

Professional athletes are often at the forefront of embracing cutting-edge technologies to optimize their performance, recover faster, and prevent injury. The integration of AI, RF therapy, and EMS into their training and recovery protocols has provided them with the tools needed to push their physical boundaries.

1. **RF and EMS for Enhanced Performance and Recovery**: Top athletes, from bodybuilders to endurance runners, are incorporating RF and EMS treatments to accelerate recovery post-training and enhance muscle growth. RF therapy, with its ability to penetrate deep muscle layers, helps stimulate collagen production and increase circulation, promoting tissue healing and reducing muscle stiffness. EMS, on the other hand, allows athletes to stimulate muscle fibers that are difficult to engage through conventional training alone, thereby strengthening muscles without the need for additional physical strain.

   A notable example is a professional basketball player recovering from a torn hamstring. After receiving a regimen of RF therapy coupled with EMS, the athlete was able to return to full activity weeks ahead of schedule, reducing downtime significantly. The combination of therapies helped not only heal the muscle but also improve muscle tone, making the player more agile and powerful than before the injury.

2. **AI-Driven Recovery Programs**: AI has revolutionized the way athletes approach training and recovery by offering highly personalized treatment plans. Through continuous data analysis—such as muscle activation, fatigue levels, and previous injuries—AI can adjust training intensity, recovery strategies, and bioelectric treatment regimens in real time. This data-driven approach ensures that athletes are recovering optimally while avoiding the risk of overtraining or injury.

   For example, a marathon runner uses AI to monitor muscle stress during long-distance runs. By analyzing real-time feedback from wearable sensors, AI suggests adjustments to the runner's recovery routine, including RF therapy sessions for muscle soreness, and EMS treatments to target specific muscle groups to prevent imbalances.

Case Studies of Personal Transformation

While professional athletes may seem like the most obvious candidates for bioelectric muscle therapies, everyday individuals are also experiencing extraordinary benefits. Case studies from various demographic groups show that these therapies can significantly improve muscle health, aid in recovery, and enhance physical well-being.

1. **Personal Transformation in a Senior Athlete**: A 65-year-old amateur triathlete had been struggling with muscle loss and recovery times due to aging. After integrating EMS therapy into his training routine, he saw substantial improvements in muscle strength and endurance. The use of EMS not only helped him rebuild muscle mass lost over the years but also enhanced his performance by improving muscle activation during training.

Additionally, combining AI-driven recovery plans with RF therapy further accelerated his muscle regeneration, allowing him to recover from long runs and swim sessions more quickly. Within three months, he achieved a personal best in his triathlon times, a feat he hadn't accomplished in years.

2. **Recovery from Chronic Injuries**: A woman in her mid-40s had been dealing with chronic back pain and muscle weakness due to an old injury. Traditional physiotherapy had yielded limited results, and she was hesitant to resort to surgery. After starting a combination of RF therapy and EMS treatments tailored to her specific injury site, she experienced significant relief from pain and discomfort.

   The RF therapy helped reduce inflammation and promote collagen production in the damaged tissue, while EMS stimulated the muscles surrounding the injured area to regain strength. After several weeks, she regained full mobility, allowing her to resume activities like yoga and hiking that she thought she would never be able to enjoy again.

Public Health Implications: The Broader Societal Benefits

The positive impact of bioelectric muscle therapies extends far beyond athletes and individual recovery stories. When viewed from a public health perspective, these technologies hold the potential to reduce the strain on healthcare systems, enhance overall physical health, and improve quality of life for the general population.

1. **Preventing Sarcopenia in the Aging Population**: As discussed in Chapter 17, sarcopenia—the age-related loss of muscle mass and function—is a major health concern in aging populations. Bioelectric muscle therapies have been shown to reverse some of the effects of sarcopenia by promoting muscle regeneration and increasing muscle strength in older adults. Regular treatments with EMS and RF therapy, combined with proper nutrition and exercise, can help preserve muscle function in seniors, allowing them to maintain independence for longer. Research into the effectiveness of EMS and RF therapy in older adults has been promising, with many individuals reporting improved muscle tone, increased strength, and a reduction in fall risk. These therapies offer a non-invasive, drug-free solution to combat muscle weakness and the physical limitations that come with aging.

2. **Reducing Healthcare Costs**: The cost of treating chronic injuries, muscle degeneration, and age-related muscle loss can be a burden on healthcare systems worldwide. By implementing bioelectric muscle therapies, such as RF and EMS, as part of rehabilitation programs, the need for expensive surgeries and long-term medication can be minimized. This could lead to significant cost savings for both individuals and healthcare providers while improving patient outcomes. Early intervention with EMS and RF therapies can help prevent the escalation of muscle conditions, allowing individuals to maintain physical function longer without requiring invasive treatments. For instance, studies suggest that integrating EMS treatments into physical therapy for patients with chronic musculoskeletal pain can reduce the need for pain medication, thus lowering healthcare costs over time.

Conclusion

The real-world applications of bioelectric muscle therapies, particularly in the fields of muscle regeneration, growth, and recovery, continue to unfold in exciting ways. Whether it's professional athletes recovering faster from injury, seniors regaining strength and independence, or individuals transforming their physical capabilities, the benefits of AI, RF, and EMS therapies are profound and far-reaching. These success stories highlight the potential of bioelectric muscle systems to change lives, improve health outcomes, and offer solutions to some of the most common physical challenges faced by people of all ages.

As these technologies evolve, they will likely become even more accessible, enhancing the ability to personalize treatments and expand their applications. The future of bioelectric muscle therapies holds the promise of not only improving physical performance but also significantly reducing the global healthcare burden by offering preventative and regenerative solutions to muscle-related issues.

Chapter 20: Building a Bioelectric Muscle Therapy Business

The rise of bioelectric muscle therapies, including AI-driven optimization, radiofrequency (RF) therapy, and electromagnetic stimulation (EMS), presents a tremendous business opportunity for those interested in the fields of health, wellness, and performance enhancement. With more people seeking advanced, non-invasive treatments for muscle growth, injury recovery, and aging-related muscle degeneration, creating a bioelectric muscle therapy business could be both financially rewarding and socially impactful. This chapter will guide you through the process of setting up a business focused on these advanced therapies, from the initial planning stages to marketing and scaling, while also addressing the legal and ethical considerations involved.

Starting a Bioelectric Muscle Clinic

Building a bioelectric muscle therapy clinic requires careful planning, research, and investment in both the technology and customer experience. Below is a step-by-step guide to launching your own clinic.

Market Research and Business Planning

- **Service offerings**: What therapies will you offer? Will it be focused on performance, rehabilitation, or anti-aging?
- **Pricing strategy**: Determine how to price your services competitively while ensuring profitability.
- **Operational plan**: Outline the day-to-day operations of the clinic, including staff requirements, hours of operation, and inventory management.
- **Financial projections**: Estimate your startup costs, expected revenue, and growth trajectory.

2. **Securing Equipment and Technology**: A critical part of your bioelectric muscle therapy business is acquiring the necessary equipment and technology. You will need to invest in high-quality RF therapy machines, EMS devices, and the software necessary to deliver AI-driven treatments. Ensure that the equipment you choose is backed by clinical evidence, safety certifications, and customer testimonials to ensure efficacy.

   Consider purchasing equipment from reputable manufacturers and suppliers with a history of providing reliable, safe devices. Additionally, integrating AI-driven software that personalizes treatment plans for clients will enhance the uniqueness of your business and differentiate you from competitors.

3. **Setting Up a Facility**: The physical space of your bioelectric muscle clinic plays a key role in the customer experience. Design a welcoming, comfortable environment with modern, clean, and private treatment rooms. You will need:

- **Treatment areas**: Equipped with RF therapy and EMS devices.
- **Reception and waiting areas**: With a calming atmosphere and amenities for clients.
- **Office space**: For administrative work, record-keeping, and client consultations.
- **Cleanliness and hygiene**: As your clinic will involve physical therapies, maintaining a sanitary and sterile environment is crucial.

Hiring Qualified Staff

- **Therapists**: Staff members experienced with RF therapy, EMS, and other physical therapy modalities. Ideally, these individuals should have certifications in physiotherapy or sports medicine.
- **AI Technicians**: For clinics that will utilize AI-driven protocols, you may need AI specialists who can ensure the software is integrated and optimized for the best patient outcomes.
- **Administrative and marketing staff**: Personnel who can manage customer service, bookings, and day-to-day operations, as well as help in digital marketing and community outreach.

Creating Treatment Protocols

- **Muscle regeneration protocol**: A tailored plan using RF and EMS to aid recovery from injuries or surgery.
- **Performance enhancement**: An AI-assisted program designed to stimulate muscle growth and improve athletic performance.
- **Anti-aging muscle therapy**: Protocols focusing on the rejuvenation of muscle mass and function in aging populations.

By combining RF and EMS therapies with AI-driven data analytics, you can ensure that each client receives a personalized and optimal treatment plan.

Marketing Bioelectric Muscle Systems

To succeed in the bioelectric muscle therapy business, effective marketing is essential to attract new clients and establish your clinic as a trusted name in the industry. Below are strategies to help you market your business:

1. **Develop an Online Presence**: A strong online presence is crucial in the modern digital world. Create a user-friendly website where potential clients can learn about your services, book appointments, and read customer reviews. Include educational content about the benefits of RF therapy, EMS, and AI-driven treatments.

   Utilize social media platforms like Instagram, Facebook, and YouTube to showcase real-world success stories, client testimonials, and educational posts. These platforms allow you to engage with clients, share valuable content, and promote offers or discounts.

2. **Educational Content and Webinars**: Educate your target audience about the science and benefits of bioelectric muscle therapies. Host webinars, publish blog posts, and create video content that explains the mechanisms of RF therapy, EMS, and how AI can optimize muscle recovery and growth. The more educated your potential clients are, the more likely they are to trust and choose your services.

3. **Partnering with Fitness Centers and Rehabilitation Clinics**: Collaborate with local gyms, fitness centers, physiotherapy clinics, and even sports teams. Offer introductory discounts or partnership packages that allow members of these organizations to try your bioelectric muscle therapies. These partnerships can help establish your clinic as an industry leader and build a steady stream of referrals.

4. **Local Advertising and Community Engagement**: Use local advertising channels such as flyers, local magazine ads, or radio spots to promote your clinic. Attend community events and health fairs to raise awareness about bioelectric therapies and introduce your clinic to a broader audience. Networking with local healthcare professionals can also help you build credibility and referral networks.

Ethics and Legal Considerations

Running a bioelectric muscle therapy clinic comes with various legal and ethical responsibilities. Below are key considerations:

1. **Licensing and Certifications**: Ensure that all therapists and staff hold the necessary certifications and licenses required by local health authorities. This is crucial to maintaining the integrity of your clinic and ensuring that you comply with healthcare regulations.

2. **Informed Consent and Privacy**: Clients must be fully informed about the procedures, benefits, and risks associated with the treatments they will receive. Obtain written consent before performing any therapy and ensure client data is securely stored to comply with privacy laws such as HIPAA in the United States.

3. **Maintaining Safety Standards**: Adhere to safety guidelines for both equipment usage and client care. Regular maintenance of all devices and equipment, along with staff training, will minimize the risk of injury during treatment sessions.

Conclusion: Scaling and Future Trends

As your bioelectric muscle therapy business grows, consider scaling your operations by opening additional locations or offering at-home services with portable equipment. You could also explore the integration of advanced wearable devices and further innovations in AI to continue personalizing treatments.

With the growing demand for non-invasive, results-driven therapies, bioelectric muscle therapy is well positioned to be a major player in the wellness and healthcare sectors. By focusing on personalized treatment, offering top-quality services, and continuously innovating, you can create a successful and impactful business that meets the needs of a diverse and ever-evolving market.

The future is bright for bioelectric muscle therapies, and your clinic could be at the forefront of this exciting revolution in health and performance.

4o mini

You said:

write chapter 21 for the book Book Title: Bioelectric Muscle: Harnessing AI, RF Therapy & Electromagnetic Stimulation for Growth and Regeneration Introduction (10 Pages) Overview of Bioelectricity: Introduction to the concept of bioelectricity, how electrical signals work in biological systems. Why Focus on Muscle Growth: The need for innovative methods for muscle regeneration, growth, and recovery. AI, RF, and Electromagnetic Stimulation: A brief overview of the core technologies—AI-driven optimization, radiofrequency therapy, and electromagnetic stimulation—used for muscle growth. Chapter 1: The Science of Bioelectricity (12 Pages) Fundamentals of Bioelectricity: What bioelectricity is and its role in cellular functions. Electric Signals in the Body: Exploration of how nerves, muscles, and cells communicate through electrical signals. Electromagnetic Fields and Biological Tissue: How RF and electromagnetic fields affect biological tissues at the cellular level. Chapter 2: Muscle Physiology and Growth (14 Pages) Muscle Structure: Anatomy of muscle fibers and the role of protein synthesis. Hypertrophy and Muscle Repair: The process of muscle growth and repair through mechanical stress and nutrient intake. Microtears and Muscle Recovery: How microtears contribute to muscle regeneration and the need for external stimuli. Chapter 3: Understanding Growth Hormone (12 Pages) What Is Growth Hormone?: Its role in muscle growth, fat metabolism, and tissue regeneration. The Endocrine System: How growth hormone is released from the pituitary gland and the signaling pathways involved.

The Anabolic Process: How growth hormone triggers muscle synthesis and helps in cellular rejuvenation.

Chapter 4: The Role of AI in Enhancing Muscle Growth (14 Pages) Artificial Intelligence and Data: The role of AI in collecting, analyzing, and optimizing data to stimulate muscle growth. AI in Personalizing Treatment: How AI can create personalized muscle growth plans based on individual physiology. AI-Driven Biofeedback: The use of AI to monitor and adjust muscle training and recovery in real time. Chapter 5: Radiofrequency (RF) Therapy Explained (16 Pages) What Is RF Therapy?: An in-depth explanation of how RF energy is used in medical treatments. Mechanism of Action: How RF waves stimulate deep tissues and encourage blood flow, collagen production, and muscle healing. RF Therapy for Muscle Regeneration: Case studies of how RF therapy has been used to improve muscle recovery and growth. Chapter 6: Electromagnetic Stimulation (EMS) and Its Applications (14 Pages) What Is EMS?: Introduction to EMS technology and its principles. How EMS Works on Muscles: The effect of electrical impulses on muscle fibers, leading to contraction and growth. EMS for Injury Recovery and Rehabilitation: How EMS is used in rehabilitation to restore muscle strength and function. Chapter 7: Combining AI with RF and EMS (14 Pages) The Synergy of AI and Electrotherapy: How AI can optimize RF and EMS treatment for better results. Real-Time Monitoring and Adjustments: The use of sensors and AI algorithms to adjust treatment parameters based on feedback. Creating Custom Protocols: How AI can help create personalized RF and EMS protocols for muscle growth and regeneration.

Chapter 8: The Bioelectric Muscle Approach to Hormonal Stimulation (12 Pages)

Stimulating Growth Hormone Release: Using electrical signals and RF therapy to trigger the release of growth hormone.

Hormonal Pathways Activation: Exploring how electrical and electromagnetic stimulation activates key hormonal pathways. Optimizing Hormonal Balance for Muscle Growth: How consistent stimulation enhances muscle regeneration by balancing hormones. Chapter 9: Integrating Nutrition with Bioelectric Stimulation (14 Pages) Role of Nutrition in Muscle Growth: The importance of protein, carbohydrates, and fats in muscle recovery and growth. Supplements and Bioelectric Stimulation: How combining supplements like amino acids and growth factors can support bioelectric muscle therapies. Fueling the Bioelectric Process: How specific nutrients enhance the bioelectric signals used for muscle regeneration. Chapter 10: RF and EMS in Clinical Practice (16 Pages) Current Applications in Medicine: How RF and EMS are being used in medical fields such as physiotherapy and rehabilitation. Case Studies: Real-world examples of individuals using RF and EMS for muscle growth, injury recovery, and performance enhancement. Safety and Efficacy: An overview of the safety standards and regulatory considerations in clinical applications of RF and EMS. Chapter 11: The Role of Sleep and Rest in Bioelectric Muscle Growth (12 Pages) The Importance of Recovery: How sleep and rest contribute to the growth process by allowing muscles to repair and grow. Hormonal Release During Sleep: Exploring how sleep optimizes hormonal pathways like growth hormone. Integrating Rest with AI-Driven Muscle Growth: How AI can monitor rest cycles to enhance recovery and muscle growth.

Chapter 12: The Future of AI and Bioelectric Muscle Growth (14 Pages)

Emerging Technologies: Exploring future innovations in AI, RF therapy, and EMS that could accelerate muscle growth.

Integration with Wearable Devices: How wearables could monitor muscle performance and adjust treatments in real-time using AI. Bioelectric Muscle in Sports and Fitness: The potential applications for athletes and fitness enthusiasts seeking optimized muscle growth and recovery. Chapter 13: Ethical Considerations in Bioelectric Muscle Therapy (14 Pages) Ethical Issues with AI in Medicine: Discussing privacy concerns, consent, and the use of AI in health-related treatments. Bioelectric Stimulation in Non-Medical Settings: Addressing the implications of using RF and EMS for performance enhancement outside clinical environments. Balancing Innovation and Ethics: How to approach the balance between cutting-edge technologies and their responsible application. Chapter 14: Building the Bioelectric Muscle System (14 Pages) Components of a Bioelectric System: An overview of the necessary hardware and software to implement AI-driven RF and EMS treatments. Designing a Personalized System: How to create a custom bioelectric muscle system tailored to individual needs. Future Trends in Bioelectric Muscle Systems: Predicting how these systems will evolve with technological advancements. Chapter 15: Overcoming Challenges in Bioelectric Muscle Therapy (14 Pages) Resistance to New Technologies: Overcoming skepticism and acceptance hurdles in adopting AI, RF, and EMS for muscle growth. Technological Barriers: Addressing challenges in sensor technology, AI integration, and RF application. Optimizing Patient Outcomes: How to maximize the success of bioelectric muscle systems for diverse populations.

Chapter 16: RF and EMS for Injury Prevention (12 Pages)

Preventing Muscle Strain and Overuse Injuries: How RF and EMS can be used as proactive measures for injury prevention.

Role of Electrical Stimulation in Tissue Repair: How RF therapy speeds up tissue healing and reduces the risk of injury. Case Examples: Documenting successful injury prevention programs using bioelectric technologies. Chapter 17: Bioelectric Muscle in Aging and Regenerative Medicine (12 Pages) How Aging Affects Muscle Regeneration: The impact of aging on muscle mass and growth potential. Using Bioelectric Muscle to Reverse Age-Related Muscle Loss: How AI, RF, and EMS can help combat sarcopenia and other age-related muscle conditions. Potential in Stem Cell Therapy: Exploring how bioelectric stimulation can be combined with regenerative techniques like stem cells. Chapter 18: Cost-Effective Solutions for Bioelectric Muscle Growth (12 Pages) Affordable Technology: How to make RF therapy and EMS more accessible to a wider audience. AI and Cloud-Based Systems: Leveraging cloud-based AI to reduce the cost of muscle growth optimization systems. Commercial and Home Use Solutions: How consumers can access and benefit from bioelectric muscle therapies at home. Chapter 19: Real-World Applications and Success Stories (14 Pages) Professional Athletes and Bioelectric Muscle Growth: How athletes use RF and EMS systems for performance enhancement and recovery. Case Studies of Personal Transformation: Stories of individuals who have achieved remarkable muscle growth or recovery using these systems. Public Health Implications: The broader societal benefits of bioelectric muscle therapy in promoting health and fitness.

Chapter 20: Building a Bioelectric Muscle Therapy Business (14 Pages) Starting a Bioelectric Muscle Clinic: A step-by-step guide on setting up a business focused on RF therapy, EMS, and AI for muscle growth.

Marketing Bioelectric Muscle Systems: How to promote bioelectric systems to a wide consumer base. Ethics and Legal Considerations: Key legal aspects to consider when running a bioelectric therapy business. Conclusion: The Future of Bioelectric Muscle (10 Pages) Summarizing the Key Concepts: Recap of the AI, RF, and EMS technologies discussed in the book. Looking Forward: Final thoughts on how this field will evolve and the potential impact on muscle regeneration and growth. Call to Action: Encouraging further research and application of these technologies to revolutionize how we approach muscle health.

ChatGPT said:

ChatGPT

Chapter 20: Building a Bioelectric Muscle Therapy Business

The growing demand for advanced treatments like radiofrequency (RF) therapy, electromagnetic stimulation (EMS), and AI-driven muscle regeneration systems creates an exciting opportunity for entrepreneurs to enter the burgeoning field of bioelectric muscle therapy. As the potential for these technologies to transform muscle growth, recovery, and regeneration becomes clearer, setting up a bioelectric muscle therapy business can be both a lucrative and impactful endeavor. In this chapter, we will explore the steps involved in building a successful bioelectric muscle therapy business, the critical components of such a system, marketing strategies, and key legal and ethical considerations.

Starting a Bioelectric Muscle Clinic: A Step-by-Step Guide

1. **Market Research and Feasibility Study**: Before diving into setting up a bioelectric muscle therapy clinic, conducting thorough market research is crucial. Assess the demand for bioelectric treatments in your region, the level of competition, and the potential client base. Understand your target audience—whether athletes, fitness enthusiasts, seniors experiencing age-related muscle degeneration, or individuals recovering from injuries.

   A feasibility study should also evaluate the technological costs, the initial investment required, and the potential for return on investment (ROI). Identify the most suitable location, whether it's a standalone clinic or a part of an existing wellness or rehabilitation center.

2. **Business Plan and Financial Projections**: A solid business plan is essential for any new venture. This plan should include:

- **Mission and vision** of your bioelectric muscle therapy business.
- **Target market and customer segmentation**.
- **Overview of the treatments and services** you will offer, including RF therapy, EMS treatments, and AI optimization programs.
- **Financial projections** for the first 3–5 years, including capital requirements, operating expenses, and expected revenue streams.
- **Marketing and sales strategies** to attract and retain clients.
- **Staffing needs** such as therapists, technicians, and AI specialists to operate the technology.

3. **Licensing and Regulatory Requirements**: Opening a therapy clinic requires navigating the regulatory landscape. Different countries and regions have varying laws governing the use of medical devices and therapies. Ensure that the equipment you plan to use, such as RF machines and EMS devices, are FDA-approved or meet the regulatory requirements in your area.

   Additionally, health and safety standards for clinics must be adhered to, including hygiene protocols and maintaining patient confidentiality (HIPAA compliance in the U.S.). It is crucial to consult with legal and health professionals to ensure your clinic operates within the legal framework.

4. **Setting Up Your Facility**: Once the necessary legal and business groundwork has been laid, it's time to focus on setting up the clinic. The space should be designed to create a comfortable, welcoming atmosphere for clients. The treatment rooms must be equipped with the latest RF and EMS machines, along with wearables and monitoring devices that allow for AI-driven personalization.

   Other essential aspects to consider include:

- Reception area and waiting room design.
- Staffing: Hiring qualified therapists or technicians trained in RF therapy, EMS, and AI-powered treatments.
- Treatment room setup: Ensure all equipment is calibrated and ready for use.

AI-Integrated Treatment Systems

Marketing Bioelectric Muscle Systems

Once your clinic is operational, the next step is to create awareness and attract clients.

Effective marketing is key to building a strong brand and establishing a client base.

1. **Building a Strong Online Presence**: In today's digital age, having a robust online presence is crucial. Create an informative, user-friendly website that explains the science behind bioelectric muscle therapies, the services offered at your clinic, and success stories. Including a blog or educational section that discusses muscle growth, recovery techniques, and the benefits of RF and EMS therapies can also help position your clinic as a thought leader in the field.

   Additionally, social media platforms like Instagram, Facebook, and YouTube can be leveraged to showcase client testimonials, before-and-after transformations, and educational content. This can build trust and inspire potential customers to explore bioelectric therapy treatments.

2. **Collaborations and Partnerships**: Partnering with local gyms, fitness centers, and wellness providers can help broaden your reach. Offering packages or referral programs where clients can access discounts on treatments when they sign up through these partnerships can drive traffic to your clinic.

   Engaging with professional athletes or fitness influencers who believe in the power of bioelectric muscle therapies can also serve as an endorsement and attract new clients.

3. **Customer Retention**: Retaining clients is equally important as attracting new ones. Offering personalized follow-up treatment plans, loyalty programs, and special offers on additional services or products (such as supplements or wearable devices) can help ensure repeat business.

   Offering a mix of physical and digital treatments can also increase client engagement. For instance, offering at-home devices for EMS therapy or app-based tracking for AI-generated muscle growth plans can keep clients connected to your clinic and the services you offer even after they leave.

Ethics and Legal Considerations

Operating a bioelectric muscle therapy business involves navigating a range of ethical and legal issues, particularly with the integration of AI and medical treatments.

1. **Informed Consent and Data Privacy**: Informed consent is a vital aspect of any medical or therapeutic treatment. Ensure that clients fully understand the procedures and technologies being used in their treatment plans. They should be informed of the benefits, risks, and alternatives before beginning any therapy. With AI-driven treatments and wearable devices that track personal data, privacy becomes an even more significant concern. Implement stringent data security measures to protect sensitive client information, ensuring compliance with data protection regulations such as GDPR (General Data Protection Regulation) or HIPAA in the U.S.

2. **Ethical Use of AI**: While AI can enhance personalized treatments, it's important to maintain ethical standards in its application. AI should never replace a medical professional's expertise but rather be used as a tool to enhance decision-making. Clinics must ensure that AI-driven systems are transparent, accountable, and free from bias in their recommendations.

3. **Ensuring Accessibility**: A significant ethical consideration in bioelectric muscle therapy is ensuring that these advanced treatments are accessible to a broader audience, including individuals with lower socioeconomic status. Offering tiered pricing, payment plans, or community outreach programs can help make these services more accessible to those in need.

Conclusion: Building a Sustainable and Impactful Bioelectric Muscle Therapy Business

Building a bioelectric muscle therapy business is an exciting and rewarding endeavor. By combining advanced technologies like AI, RF therapy, and EMS, you can provide clients with groundbreaking solutions for muscle growth, recovery, and regeneration. The demand for personalized, non-invasive therapies is only expected to grow, and establishing a clinic in this emerging field offers significant potential for both business success and making a positive impact on people's lives.

To ensure long-term success, focus on integrating state-of-the-art technologies, fostering strong customer relationships, and adhering to ethical and legal standards. By doing so, you will create a business that not only thrives in the competitive health and wellness market but also helps revolutionize how people approach muscle health and recovery.

Chapter 22: Summary, Future Directions, and Call to Action

In the previous chapters, we have journeyed through the fascinating world of bioelectric muscle therapy, exploring the roles of artificial intelligence (AI), radiofrequency (RF) therapy, and electromagnetic stimulation (EMS) in promoting muscle growth, recovery, and regeneration. We've delved into the science of bioelectricity, the physiological processes involved in muscle growth, and how cutting-edge technologies can transform how we approach muscle health.

In this final chapter, we will recap the key concepts discussed, look toward the future of bioelectric muscle therapy, and present a call to action for further research and the widespread application of these transformative technologies.

Recap of Key Concepts

1. **Bioelectricity and Its Role in Muscle Growth**: Bioelectricity—the study of electrical signals in biological systems—is central to understanding how muscles grow, repair, and regenerate. Electrical impulses are essential for cellular communication, and in the case of muscle cells, bioelectric signals play a significant role in stimulating muscle contractions, initiating repair, and promoting growth. By harnessing these electrical signals with advanced technologies, we can optimize muscle regeneration and recovery.

2. **AI-Driven Personalization of Muscle Therapy**: Artificial intelligence has emerged as a key player in enhancing muscle growth and recovery. Through the collection and analysis of data, AI allows for personalized treatment plans that are tailored to the individual's unique physiology. AI-driven systems can adapt to the real-time needs of the body, optimizing the combination of RF therapy, EMS, and other interventions for maximum effectiveness.

3. **Radiofrequency (RF) Therapy**: RF therapy uses energy to stimulate deep tissues, enhancing blood flow, collagen production, and muscle healing. This non-invasive technique accelerates muscle recovery and can be particularly beneficial for individuals recovering from injuries or those seeking to enhance their muscle regeneration post-workout. We have seen how RF therapy has proven effective in clinical settings for both muscle growth and rehabilitation.

4. **Electromagnetic Stimulation (EMS)**: EMS works by using electrical impulses to simulate muscle contractions, promoting muscle strength, recovery, and growth. It has become an invaluable tool in rehabilitation settings, enabling faster recovery and muscle function restoration after injuries. EMS has also gained traction in fitness circles, helping individuals achieve optimized muscle performance.

5. **Combining AI with RF and EMS**: The synergy between AI and electrotherapy (RF and EMS) offers a groundbreaking approach to muscle growth. AI optimizes the delivery of RF and EMS treatments by adjusting parameters based on feedback from sensors, ensuring that each session is personalized to the client's specific needs. This fusion of technologies ensures that bioelectric muscle therapy is both effective and safe.

6. **Nutrition and Hormonal Stimulation**: Muscle growth does not happen in isolation. Nutrition plays a pivotal role in fueling muscle regeneration. The right balance of macronutrients, amino acids, and supplements can support bioelectric signals and enhance the effects of RF and EMS therapies. Additionally, the release of growth hormones, triggered by both electrical stimulation and natural processes like sleep, further accelerates muscle repair and growth.

7. **Cost-Effective Solutions for Bioelectric Muscle Therapy**: One of the challenges in making these advanced therapies accessible to a wider audience is the cost of technology. However, by leveraging cloud-based AI and providing at-home solutions for EMS and RF therapy, these treatments can be made more affordable and accessible, empowering individuals to take control of their muscle health in the comfort of their own homes.

8. **Real-World Applications and Success Stories**: From professional athletes to everyday individuals seeking enhanced performance or recovery, bioelectric muscle therapy has proven its worth. Case studies have highlighted the successful integration of RF and EMS technologies in improving muscle strength, accelerating recovery, and reversing age-related muscle loss. These success stories underscore the profound impact that bioelectric muscle therapy can have on people's lives.

Future Directions

The future of bioelectric muscle therapy is incredibly promising. As technologies evolve, we can expect significant advancements in the way we approach muscle growth and regeneration. Here are a few key areas to watch:

1. **Advancements in AI and Machine Learning**: As AI and machine learning continue to progress, personalized muscle therapy will become even more precise. AI systems will not only be able to optimize RF and EMS treatment protocols but will also be able to predict individual muscle needs, preventing injuries and maximizing recovery potential.

2. **Integration with Wearable Devices**: Wearables that monitor muscle activity, fatigue levels, and recovery metrics will be integrated with bioelectric muscle systems. These devices will provide real-time feedback and adjustments to RF and EMS therapy sessions, creating a truly personalized and responsive treatment experience.

3. **Regenerative Medicine**: Bioelectric muscle therapy has the potential to intersect with other regenerative medicine techniques, such as stem cell therapy. By combining these therapies, we may be able to accelerate the body's natural healing processes, reversing muscle degeneration and restoring muscle tissue to a more youthful state.

4. **Broader Clinical Adoption**: As research continues to validate the efficacy of RF, EMS, and AI-driven therapies, we can expect these treatments to become more common in clinical settings. The use of bioelectric therapies in rehabilitation centers, sports medicine, and even for aging populations will become widespread, providing innovative solutions for muscle health across various demographics.

5. **Commercialization and Accessibility**: With technological advancements, the cost of these therapies will likely decrease, making them more accessible to the general public. Commercial and home-use solutions, including affordable RF and EMS devices, will help democratize access to muscle growth and recovery therapies.

Call to Action

The innovations discussed in this book represent just the beginning of a revolution in muscle therapy. To truly realize the potential of bioelectric muscle therapy, there are several key steps that both professionals and individuals can take:

1. **Invest in Research**: Further research is needed to explore the long-term effects of RF and EMS therapy, as well as their combined impact with AI. Encouraging academic and clinical studies will provide the data needed to refine and perfect these therapies.

2. **Educate the Public**: As the benefits of bioelectric muscle therapy become more widely understood, it is crucial to educate the public on its advantages. Clinics, gyms, and wellness centers should take the initiative in providing information and access to these transformative technologies.

3. **Explore New Applications**: While bioelectric muscle therapy is already showing promise in injury recovery, muscle regeneration, and fitness enhancement, there are many other potential applications. Research into how these therapies can help with neurological conditions, chronic pain, and aging-related muscle loss should be encouraged.

4. **Develop Affordable Solutions**: For bioelectric muscle therapy to become a mainstream treatment, cost-effective solutions must be developed. This includes making home-use devices more affordable and accessible, leveraging cloud-based AI systems, and reducing the overall cost of RF and EMS treatments in clinical settings.

5. **Commit to Ethical Practices**: As with any new technology, ethical considerations must be at the forefront of bioelectric muscle therapy's integration into medicine. Privacy concerns, informed consent, and responsible AI practices must be prioritized to ensure that these technologies are used for the betterment of society.

Conclusion

The field of bioelectric muscle therapy is poised to redefine how we think about muscle growth, recovery, and regeneration. With the combination of AI, RF therapy, and EMS, we have the opportunity to accelerate muscle healing, prevent injuries, and optimize performance like never before. By continuing to push the boundaries of these technologies and making them accessible to a wider audience, we can truly revolutionize how we approach muscle health.

Now is the time to embrace these innovations and take action—whether by starting a bioelectric therapy clinic, investing in research, or simply exploring these therapies for personal growth and recovery. The future of muscle health is bioelectric, and the possibilities are limitless.

Chapter 23: The Long-Term Impact of Bioelectric Muscle Therapy on Global Health

The integration of bioelectric muscle therapy into both medical practices and personal wellness routines presents an opportunity to radically improve global health outcomes. As we have explored in this book, technologies such as AI-driven optimization, RF therapy, and EMS are already proving their potential to enhance muscle growth, recovery, and regeneration. This chapter will delve into the broader societal impact of these technologies, looking at how they could shape the future of healthcare, athletic performance, aging populations, and overall health across the globe.

Transforming Healthcare Systems

Bioelectric muscle therapy holds great promise in revolutionizing healthcare by offering non-invasive treatments that are both effective and cost-efficient. Traditional therapies for muscle recovery, pain management, and rehabilitation often require surgical interventions, extended hospital stays, and significant pharmaceutical usage. In contrast, RF and EMS technologies allow patients to recover in a non-invasive manner, reducing risks associated with surgery and minimizing the need for drugs.

1. **Enhanced Rehabilitation**: As the world's population ages, the prevalence of musculoskeletal disorders and muscle loss due to aging (sarcopenia) will increase. By integrating bioelectric muscle therapy into rehabilitation protocols, healthcare providers can offer better recovery times and more efficient healing processes. EMS, for example, can be used post-surgery or injury to stimulate muscle fibers and promote healing, reducing recovery times and improving the quality of life for patients.

2. **Chronic Disease Management**: Many chronic diseases, including arthritis, diabetes, and heart disease, can lead to muscle atrophy, loss of muscle function, and overall weakness. Integrating bioelectric therapies into the treatment of these conditions could help patients maintain muscle mass, increase strength, and improve overall mobility. This could reduce the burden on healthcare systems by lowering the need for long-term care and hospitalization.

3. **Reducing the Need for Pharmaceuticals**: By offering an alternative to pain management, particularly for musculoskeletal pain, RF and EMS therapies could reduce the dependency on opioid medications and other painkillers, addressing a growing global health crisis related to the overuse of pharmaceuticals.

Revolutionizing Sports and Athletic Performance

The field of sports medicine stands to benefit greatly from bioelectric muscle therapy. Athletes of all levels are always seeking ways to enhance performance, speed up recovery, and prevent injuries. Bioelectric muscle therapy technologies are already being used by elite athletes, and their impact is becoming more pronounced as their benefits become widely recognized.

1. **Optimizing Performance**: Athletes can use AI-driven bioelectric muscle systems to optimize training by providing real-time feedback on muscle performance, fatigue, and recovery. AI can personalize training and recovery protocols, ensuring that athletes train at their peak potential while avoiding overtraining.
2. **Injury Prevention and Recovery**: Preventing injuries in sports is a constant challenge. RF and EMS therapies can be used proactively to reduce muscle strain and overuse injuries. With the ability to stimulate muscles in a targeted and controlled manner, athletes can strengthen muscles before intense physical activity, enhancing muscle resilience and reducing the likelihood of injuries.
3. **Expedited Recovery**: Recovery times can be drastically reduced by using bioelectric therapies to stimulate the muscles and increase blood flow to the injured area. This enhanced recovery allows athletes to return to peak performance levels more quickly, making bioelectric muscle therapy a key tool for professional sports teams and individual athletes alike.

Combating Age-Related Muscle Degeneration

The aging population presents a significant challenge to modern healthcare systems. As people age, they experience muscle loss, joint deterioration, and a decline in overall mobility. Bioelectric muscle therapy offers promising solutions to combat these age-related issues.

1. **Sarcopenia Treatment**: Sarcopenia, the age-related loss of muscle mass and strength, is a growing concern for older adults. Bioelectric muscle therapy, including RF and EMS, has been shown to help regenerate muscle tissue and maintain muscle function, providing a potential solution to this issue. By maintaining muscle mass, older adults can stay more mobile, reduce their risk of falls, and retain independence.

2. **Improved Quality of Life**: Beyond the physical benefits, bioelectric therapies can significantly improve the overall quality of life for older adults. These treatments help to alleviate chronic pain, increase mobility, and enhance strength, all of which are critical to maintaining independence as people age. Moreover, by reducing the need for surgical interventions and medications, bioelectric muscle therapies can offer a less invasive and more accessible solution for older populations.

3. **Extended Lifespan and Vitality**: While aging is inevitable, the effects of aging on the body can be mitigated. Bioelectric therapies can slow the progression of muscle degeneration, allowing individuals to maintain a more youthful level of function well into their later years. The promise of extending vitality through the use of these technologies could fundamentally change how we age, moving the focus from disease treatment to disease prevention.

Global Health and Public Policy Implications

On a broader scale, the integration of bioelectric muscle therapy could have profound effects on global health systems. The shift towards preventative healthcare and non-invasive treatments could help reduce the financial burden on healthcare systems worldwide.

1. **Accessibility and Affordability**: One of the main barriers to widespread adoption of advanced medical treatments is cost. However, with the rise of cloud-based AI systems and home-use solutions for RF and EMS therapies, bioelectric muscle treatments could become more affordable and accessible to a global population. This would allow individuals in underserved regions to benefit from these therapies, helping bridge the healthcare disparity gap.

2. **Public Health Initiatives**: Governments and public health organizations can use bioelectric muscle therapies as part of broader health initiatives aimed at improving mobility and preventing injury. For example, elderly populations could be provided access to these therapies to improve their health outcomes, reducing long-term healthcare costs associated with falls and related injuries.

3. **Promoting Wellness and Prevention**: Bioelectric muscle therapy can play a critical role in promoting wellness and preventing chronic diseases, which are on the rise globally. By focusing on muscle strength, recovery, and regeneration as part of public health strategies, societies can reduce the incidence of lifestyle diseases such as obesity, diabetes, and cardiovascular diseases.

The Path Forward: Bioelectric Muscle Therapy for All

Bioelectric muscle therapy offers transformative potential not only for individuals but also for society at large. As we look toward the future, the widespread application of these technologies could lead to healthier, longer lives, enhanced athletic performance, and a more sustainable healthcare system.

The future of bioelectric muscle therapy is incredibly bright, but it requires continued investment in research, education, and the development of affordable solutions. By integrating AI with RF and EMS therapies, expanding access to these treatments, and creating more widespread understanding of their benefits, bioelectric muscle therapy can become a cornerstone of health and wellness worldwide.

As we continue to innovate and push the boundaries of these technologies, it is crucial for all stakeholders—including healthcare providers, policymakers, athletes, and individuals—to collaborate and commit to a future where bioelectric muscle therapy becomes an integral part of our global health ecosystem.

Chapter 24: The Future of Bioelectric Muscle Therapy and Global Health

As we look toward the future, the field of bioelectric muscle therapy stands on the precipice of a healthcare revolution. Technologies such as AI-driven optimization, RF therapy, and EMS have already proven to be powerful tools for enhancing muscle growth, recovery, and regeneration. These advancements are not only changing the way we approach individual muscle health but are also setting the stage for a paradigm shift in global health systems.

This chapter explores the long-term impact of bioelectric muscle therapy on the global health landscape, the future of research and development, and how this technology can be harnessed to improve the health of individuals, communities, and populations worldwide.

Transforming Global Health Outcomes

The integration of bioelectric muscle therapy into mainstream healthcare has the potential to revolutionize the treatment of a wide array of health conditions, particularly those related to muscle and joint health. By enhancing muscle recovery, increasing muscle mass, and improving overall mobility, these therapies offer a promising solution for managing the challenges presented by aging populations, chronic diseases, and injury rehabilitation.

1. **Managing Chronic Health Conditions**: With a rise in chronic conditions like diabetes, cardiovascular disease, and obesity, which often lead to muscle loss and weakness, bioelectric muscle therapies present an opportunity to mitigate these conditions. For example, RF and EMS therapies can be integrated into treatment plans for patients with diabetes, helping them maintain muscle strength and improving their quality of life by reducing the risk of muscle atrophy and related complications.

2. **Aging Populations and Sarcopenia**: As the global population ages, sarcopenia—the gradual loss of muscle mass—has become a pressing public health issue. The implementation of bioelectric muscle therapies can help slow or even reverse muscle loss, allowing older adults to remain more active and independent for longer periods. This could dramatically reduce the societal and economic burdens associated with age-related muscle degeneration.

3. **Preventive Healthcare**: One of the most profound implications of bioelectric muscle therapy is its potential to shift the focus of healthcare from treatment to prevention. By using AI to create personalized muscle growth and regeneration protocols, individuals can proactively manage their muscle health, preventing the onset of debilitating conditions such as osteoarthritis, osteoporosis, and chronic back pain.

Future Innovations in Bioelectric Muscle Therapy

While current advancements in AI, RF, and EMS technologies have shown impressive results, the future promises even more transformative developments. Emerging technologies in this field will not only improve the precision and efficiency of bioelectric muscle therapies but also increase their accessibility and affordability, ensuring that more individuals around the world can benefit from these innovations.

1. **Integration with Wearables**: As wearable technology becomes more advanced, we can expect to see real-time bioelectric muscle therapy integrated into these devices. Imagine a smart wearable that monitors muscle performance, tracks recovery, and uses AI algorithms to adjust EMS therapy on-the-go. This would enable individuals to receive personalized, adaptive treatments without the need for a clinic visit, creating a more accessible and efficient healthcare model.

2. **AI-Driven Personalization**: Future AI systems will be able to gather and analyze vast amounts of data, including genetic, physiological, and environmental factors, to create even more tailored treatment plans. These personalized protocols will ensure that each individual receives the most effective combination of RF and EMS therapies, boosting muscle growth and regeneration while minimizing the risk of injury.

3. **Expanded Applications in Regenerative Medicine**: Beyond muscle recovery and growth, bioelectric stimulation is already being explored as a tool in regenerative medicine. The ability to stimulate stem cells and encourage tissue regeneration through electrical fields could open new doors in the treatment of spinal cord injuries, neurological disorders, and even organ regeneration. The future of bioelectric muscle therapy lies not only in enhancing muscle health but also in supporting the healing of other tissues and organs in the body.

Impact on Public Health Policy

As bioelectric muscle therapy continues to gain traction, its widespread adoption could have profound implications for public health policy and healthcare systems worldwide. Governments and healthcare organizations will need to adapt and integrate these technologies into national healthcare strategies, ensuring equitable access for all individuals, regardless of their socioeconomic status or geographic location.

1. **Equitable Access**: One of the major hurdles in healthcare is ensuring that innovative treatments are accessible to everyone. Bioelectric muscle therapy could become more widely available through the use of cloud-based systems and home-use devices, which would lower treatment costs and make these therapies more accessible to underserved populations.

2. **Regulatory Standards**: As bioelectric muscle therapies evolve, regulatory bodies will need to develop and implement standards to ensure the safety and efficacy of these treatments. This will involve creating guidelines for the use of AI in healthcare, monitoring the development of wearable devices, and ensuring that clinical applications of RF and EMS therapies meet the necessary medical standards.

3. **Health Education and Awareness**: For bioelectric muscle therapy to reach its full potential, public health campaigns and educational initiatives will be necessary to raise awareness about the benefits of these technologies. This includes educating both healthcare providers and patients on the advantages of incorporating bioelectric muscle therapy into muscle regeneration and recovery routines.

Global Wellness and Fitness Industry

The fitness industry is already beginning to embrace bioelectric muscle therapies as part of their offerings to enhance muscle performance, recovery, and regeneration. With the increasing popularity of personalized fitness programs, these technologies will likely play a central role in helping individuals achieve their health and fitness goals more efficiently.

1. **Customized Fitness Programs**: Fitness enthusiasts can use AI-driven systems to create personalized muscle-building routines that incorporate RF and EMS therapies. These treatments will help optimize muscle performance, ensuring that individuals can maximize their workouts and achieve faster, more effective results.

2. **Rehabilitation and Injury Prevention**: Bioelectric muscle therapies can play an important role in the rehabilitation of athletes and fitness enthusiasts recovering from injuries. By using EMS and RF systems, individuals can speed up recovery, prevent further injury, and maintain their training routines without the setbacks caused by muscle injuries.

3. **Proactive Health**: Beyond recovery and rehabilitation, bioelectric muscle therapies will become an essential part of a proactive health strategy. As more people embrace the concept of preventative healthcare, integrating RF and EMS therapies into daily routines will help individuals maintain muscle strength, flexibility, and vitality, ultimately contributing to long-term wellness.

Looking to the Future

The potential of bioelectric muscle therapy to improve individual and global health is vast. As technology advances, we will see more widespread adoption of AI, RF, and EMS in clinical practices, fitness routines, and home health systems. Bioelectric muscle therapy will not only help individuals achieve optimal muscle health but also serve as a cornerstone of modern preventive healthcare strategies, aging management, and rehabilitation.

The future is bright, and the journey toward making bioelectric muscle therapy accessible and effective for all is just beginning. The continued development and integration of these technologies into healthcare and wellness will transform how we approach muscle regeneration and overall health, ushering in a new era of personalized and preventative medicine.

The call to action is clear: further research, development, and public engagement are essential to fully harness the potential of bioelectric muscle therapy and realize its transformative impact on global health.

Chapter 25: The Integration of Bioelectric Muscle Therapy into Holistic Health Systems

As bioelectric muscle therapy continues to evolve, it becomes clear that its integration into broader healthcare systems holds the potential to transform not only muscle recovery and growth but also how we approach overall health and wellness. This chapter explores how these technologies can be seamlessly incorporated into holistic health systems, addressing a variety of wellness concerns, improving quality of life, and supporting the sustainability of healthcare services globally.

The Shift Towards Holistic Health Models

Historically, healthcare has largely focused on treating isolated conditions. However, as the understanding of health continues to evolve, the focus is shifting towards holistic health models—systems that take a comprehensive view of an individual's physical, mental, and emotional well-being. Bioelectric muscle therapies, particularly when combined with AI, RF therapy, and EMS, are uniquely positioned to play a pivotal role in this transition. These therapies not only address muscle growth and regeneration but also align with the broader objectives of holistic health systems that prioritize prevention, personalized care, and continuous monitoring.

1. **Preventative Health**: One of the most important aspects of holistic health is its focus on prevention. By incorporating bioelectric muscle therapy into regular wellness routines, individuals can take proactive steps to maintain muscle strength, reduce the risk of injury, and stave off conditions related to muscle loss. These preventive measures are especially valuable as they can mitigate the onset of chronic conditions like sarcopenia, osteoporosis, and arthritis, which are common in aging populations.
2. **Personalized Treatment Plans**: A key feature of holistic health models is personalization. Bioelectric muscle therapies, supported by AI-driven algorithms, allow for highly customized treatment plans based on an individual's specific needs, genetics, and health conditions. AI can analyze vast amounts of data, from sleep patterns to hormone levels, and adjust muscle regeneration protocols in real-time, optimizing results for each user. This level of personalization enhances the effectiveness of treatments and ensures that patients receive the most suitable and efficient care possible.

3. **Integrated Care Teams**: Holistic health systems often feature integrated care teams that collaborate across specialties to treat the whole person. The combination of bioelectric therapies with other medical treatments, such as physical therapy, nutritional counseling, and mental health support, creates a comprehensive approach to muscle health. For example, a patient recovering from a major surgery could receive a tailored program involving AI-optimized RF therapy and EMS treatments, coupled with nutritional guidance to enhance protein synthesis, and psychological support to address recovery-related stress.

Bioelectric Therapy and Mental Health

While the primary focus of bioelectric muscle therapy has been on physical health, there is an increasing recognition of its potential benefits for mental health as well. Muscular health and mental well-being are deeply intertwined, and the use of RF therapy, EMS, and AI to optimize muscle growth and recovery may have secondary effects that support mental health.

1. **Reducing Stress and Anxiety**: Physical activity and muscle engagement have been shown to release endorphins, which are natural mood lifters. RF and EMS therapies, particularly when used in combination with exercise or physical therapy, can help regulate mood by stimulating muscle activity and enhancing blood flow, leading to better overall physical and emotional resilience.

2. **Improving Sleep Quality**: Sleep plays a critical role in both mental and physical health, particularly in muscle recovery. By using bioelectric muscle therapies to promote deeper, more restorative sleep, individuals can experience improvements not only in muscle regeneration but also in their cognitive function and emotional well-being. AI systems that monitor sleep patterns and integrate muscle therapies into rest cycles can further enhance these benefits.

3. **Combating Depression through Physical Engagement**: For individuals suffering from depression or chronic stress, bioelectric therapies offer a non-invasive way to engage the body and mind in a healing process. The use of EMS to stimulate muscle contraction and the release of hormones like endorphins and growth hormone can alleviate symptoms of depression and contribute to a more balanced emotional state.

Collaborating with Traditional Medical Systems

The future of bioelectric muscle therapy in holistic health systems involves not only integrating these technologies into wellness and fitness routines but also collaborating with traditional healthcare systems. For bioelectric muscle therapy to be accepted widely, it must be shown to complement, rather than replace, traditional forms of treatment. This collaborative approach can be especially effective in managing chronic conditions and improving recovery times for patients undergoing surgery or rehabilitation.

1. **Post-Surgery Recovery**: One area where bioelectric muscle therapy can significantly impact holistic healthcare is post-surgery recovery. After surgical procedures, particularly those that involve muscle or joint repair, patients often face long recovery times and limited mobility. Bioelectric muscle therapies can speed up recovery by stimulating muscle regrowth, improving blood flow, and preventing muscle atrophy, allowing patients to regain strength and mobility more quickly. By integrating these therapies with traditional recovery protocols, healthcare providers can offer a more comprehensive, faster rehabilitation experience.
2. **Chronic Pain Management**: For patients suffering from chronic pain, particularly related to musculoskeletal disorders, bioelectric muscle therapy can offer non-pharmaceutical solutions to pain relief. The application of RF and EMS treatments can reduce inflammation, relieve muscle tension, and promote tissue healing, offering an effective alternative or complement to medications and invasive treatments like surgery.

3. **Muscle Health as Part of General Health Screenings**: As healthcare systems move towards more preventative models, the importance of muscle health in overall wellness cannot be overstated. Bioelectric muscle therapy systems can be integrated into general health screenings, allowing for early detection of muscle degeneration or other related conditions. Regular assessments using AI-driven systems to track muscle strength and regeneration rates could allow healthcare providers to intervene early and customize prevention strategies for at-risk patients.

The Role of Education and Awareness

For bioelectric muscle therapy to achieve its full potential within holistic health systems, widespread education and awareness are essential. Health professionals, patients, and the general public must be informed about the benefits, risks, and applications of these technologies.

1. **Training Healthcare Providers**: Medical practitioners, physical therapists, and other healthcare professionals need to be educated on the proper use of bioelectric muscle therapy tools. Understanding the science behind RF therapy, EMS, and AI optimization is critical for safely integrating these technologies into patient care. Professional training programs and certifications can ensure that these systems are used effectively, minimizing risks and optimizing outcomes for patients.

2. **Public Education Campaigns**: Public health campaigns can raise awareness about the importance of muscle health, how bioelectric therapies can support regeneration, and the benefits of incorporating these treatments into preventive care. By educating people about the role of muscle health in overall well-being, these technologies can become a regular part of health and fitness regimens for individuals of all ages.

3. **Consumer Accessibility**: As bioelectric muscle therapy becomes more mainstream, making it accessible to the public—particularly in home-use formats—will be key. Overcoming barriers to entry, such as cost, complexity, and lack of understanding, will be essential to ensuring that people worldwide can benefit from these technologies. Efforts should focus on providing affordable, easy-to-use systems and resources that guide individuals in using these therapies to enhance their health.

Conclusion: The Vision for a Healthier Future

The integration of bioelectric muscle therapy into holistic health systems offers transformative potential for individuals and communities worldwide. By combining cutting-edge technologies like AI, RF therapy, and EMS with a focus on personalized, preventative care, healthcare systems can shift from disease treatment to a more proactive, wellness-oriented approach.

As these technologies continue to develop and gain acceptance, we can expect to see a future where muscle health is prioritized alongside cardiovascular health, mental well-being, and overall vitality. The goal is not just to repair muscles but to enhance and maintain them for life, ensuring that individuals can live healthier, longer, and more fulfilling lives.

The collaboration between bioelectric muscle therapies and traditional healthcare systems will usher in a new era of comprehensive care, where advanced technology and a holistic view of health work together to support long-term wellness and quality of life. The journey towards a healthier future begins now, with bioelectric muscle therapy playing a central role in reshaping how we care for our bodies and minds.